American Forage & Grassland Council Annual Conference 2018

Forages: Opportunities for the Next Generation

Louisville, Kentucky, USA
14 - 17 January 2018

ISBN: 978-1-5108-6756-7

Printed from e-media with permission by:

Curran Associates, Inc.
57 Morehouse Lane
Red Hook, NY 12571

Some format issues inherent in the e-media version may also appear in this print version.

Printed by Curran Associates, Inc. (2018)

For permission requests, please contact American Forage and Grassland Council (AFGC)
at the address below.

American Forage and Grassland Council (AFGC)
PO Box 867
Berea, KY 40403
USA

Phone: 800.944.2342

www.afgc.org

Additional copies of this publication are available from:

Curran Associates, Inc.
57 Morehouse Lane
Red Hook, NY 12571 USA
Phone: 845-758-0400
Fax: 845-758-2633
Email: curran@proceedings.com
Web: www.proceedings.com

TABLE OF CONTENTS

AMERICAN FORAGE AND GRASSLAND COUNCIL
BYLAWS

ARTICLE I
Name

The name of this organization shall be American Forage and Grassland Council, Inc. (AFGC).

ARTICLE II
Objectives

Section A Objectives of the Council

1. To promote profitable use of forage as a prime feed resource from tilled and range land for efficient production of livestock.
2. To provide a forum for all who have a mutual interest in forage production, evaluation, and/or utilization to exchange ideas, philosophies, and research findings; and to provide a means whereby such professional contributions may be published and contributors recognized.
3. To collect, analyze, and disseminate forage technological and economic information.
4. To encourage industries serving agriculture to provide the best available products for optimum production, evaluation, and utilization of forages.
5. To identify research needs and to encourage expanded and intensified research relative to production, marketing, and utilization of forage.
6. To promote the value of forages, grazing lands, and other grasslands as soil and water conserving crops and for the abatement of pollution for the short- and long-term benefits to agriculture and environmental quality.
7. To cooperate with other organizations promoting modern technologies of forages and grassland agriculture and encourage joint programs when desirable for the most effective results.
8. To provide leadership in international forages, grazing lands, and other grassland activities.
9. All activities will be conducted to be consistent with 501c3 IRS Regulations.

ARTICLE III
Membership and Eligibility

Section A Membership Classes

1. There shall be seven classes of membership:
   a. Individual
   b. Sustaining
   c. Life membership

d. Corporate
e. Professional societies and institutions
f. Institutes, foundations and trade associations
g. State, regional, and provincial affiliate councils

2. The Board of Directors may from time to time create additional classes of members, prescribe the rules of eligibility, and define the privileges thereof.
3. There shall be three "sectors" of membership:
   a. Public – members who work for public agencies, foundations and institutes.
   b. Industry – members who work for private industries operated for profit.
   c. Producer – members whose primary source of income is from owning and/or operating a farm.

Section B Membership Eligibility

1. Individual members shall consist of any individuals interested in the objectives of this Council and individual members of affiliated state, regional, and provincial member councils in good standing.
2. Sustaining members shall consist of individual members who provide additional financial support for council objectives.
3. Life members shall consist of members who have contributed at the sustaining level for an equivalent of ten years.
4. Corporate members may consist of any agricultural, industrial, or other business firm interested in the objectives of this Council.
5. Professional societies and institution members may consist of any public or private institution, agency commission, or scientific or professional society interested in the objectives of this Council.
6. Institutes, foundations and trade associations may consist of any public or private association interested in the objectives of this Council.
7. State, regional, and provincial member councils shall consist only of councils that are officially affiliated with AFGC.
   a. Prior to being approved by the Board of Directors for official affiliation, each applying member council must submit a formal membership application, as well as information indicating geographical area included, current officers, and a general program of activity.
   b. Each member council shall have full control and responsibility over its own program and fiscal affairs. Member councils that use the umbrella AFGC 501c3 exempt status shall comply with 501c3 regulations.
   c. All individual members of affiliated state, regional, and provincial councils must pay dues to AFGC (Article III, Section E).

Section C Membership Status

1. A member shall be considered in good standing providing current fiscal year dues have been paid as set by action of the Board of Directors.
2. Each professional society, corporate member, or institute-trade association shall annually

designate one of their members as an AFGC representative. This representative shall have the rights of a member in good standing (ARTICLE III, Sections C and D). The representative shall coordinate activities and communications between AFGC and his organization.

Section D Membership Privileges

1. A member in good standing may hold office, serve on the Board of Directors, be a member of one or more committees, and attend and vote at Annual Business Meetings.

Section E Dues and Assessments

1. The Council has authority to collect from all membership categories dues and assessments for the purpose of financing the activities of the Council. A schedule of such dues and assessments for each membership category shall be set by the Board of Directors. Any changes in the annual dues and assessments schedule shall be announced by the Board of Directors at least four (4) months prior to the start of the fiscal year in which they are to become effective.

ARTICLE IV
Organization

Section A Directors

1. The directors of this organization shall consist of the Board members elected as set forth below: the immediate Past President, President, President Elect, Senior Vice President, Vice President, and Affiliate Council Advisory Committee President.
2. The President of the Council shall be Chairman of the Board of Directors.
3. There shall be twelve directors elected; four from each of the public, industry, and producer sectors.

Section B Officers and Executive Director

1. The officers shall consist of the President, President Elect, Senior Vice President, Vice President, and Treasurer.
2. The President, President Elect, Senior Vice President, and Vice President shall each hold office in ascending order from the conclusion of the annual conference to the conclusion of the next annual conference except in cases where annual meetings are less than nine months apart. Under this circumstance, the terms of all officers may be extended until the subsequent annual meeting by a majority board vote.
3. The Vice President shall be elected annually.
4. The Treasurer shall be annually appointed from the Board of Directors by the President.
5. The Executive Director shall be retained under a written contract executed by the Board of Directors. The position can be internal staff directly employed by the association or provided by a management company hired by the association to provide staff and

management services. The Executive Director shall be an ex-officio (non-voting) member of the Board of Directors, the Executive Committee, and all standing and ad hoc committees.

Section C Committees

1. All committees and representatives other than those specifically named in the Bylaws will be appointed by the President with the counsel of the Executive Committee and Board of Directors. Each individual appointed by the President to represent AFGC must be a member in good standing. Each Board member shall serve on at least one standing committee.
2. The Chairperson or Co-chairpersons of each Committee shall be a member of the Board of Directors or a current or past member of a standing committee. Otherwise his/her appointment must be authorized by the Board.
3. There shall be an Executive Committee and Nomination Committee.

a. Executive Committee
The Executive Committee shall be comprised of the Immediate Past President, President, President Elect, Senior Vice President, Vice President, and Treasurer. These people shall also serve as Board members. The Executive Director shall serve as ex-officio (non-voting) members of the Executive Committee.

b. Nomination Committee
(1) To be composed of the four most recent Past Presidents with the Immediate Past President serving as Chairperson.
(2) To annually select a nominee for Vice President. It is recommended that the Vice President be elected in rotation from each of the three sectors.
(3) To annually prepare a slate of nominees for the Board of Directors representing each of the three sectors: public, industry, and producer. At least three individuals shall be nominated for each sector, from which two will be elected.
(4) The Nomination Committee shall identify nominees so that all fields of forage and grassland interest may be represented on the Board of Directors.

c. Affiliate Council Liaison
(1) Responsibilities are to advise the Board on matters pertaining to Affiliate Councils' concerns and activities.
(2) The Affiliate Council Liaison shall be the President-Elect on the AFGC Board of Directors.

Section D Fiscal Year

1. The Council shall be fiscally operated on a July 1st to June 30th year basis.

ARTICLE V
Election of Officers and Directors

Section A Each member of the Board of Directors shall be elected for a two-year term. A Board member may serve no more than three consecutive terms.

Section B The Vice President shall be elected annually by the Board of Directors by electronic or postal ballot prior to the conference Board Meeting. Two individuals will be nominated from the same sector with a three-year rotation of sectors (public, producer, industry).

Section C The election of Directors shall be accomplished by electronic or mail ballot sent to all eligible members at least 30 days prior to the date of the annual election, the date to be set by the Executive Committee.

Section D Any Director failing to attend two consecutive Board meetings, without Board-excused absence, shall be considered resigned from the Board.

Section E In the case of the retirement, death, or resignation of an officer or a member of the Board of Directors, the vacancy shall be filled by appointment by the President of the Council, subject to confirmation by the Board of Directors. If the President is unable to act, the vacancy shall be filled by a vote of the Board of Directors.

Section F The terms of newly elected officers and directors shall begin at the conclusion of the annual conference subject to terms in Article IV Section B.

ARTICLE VI
Duties of Officials

Section A The Board of Directors shall be responsible for formulating and executing all policies of the Council in accordance with the stated objectives or such other direction as will best serve the interest of the Council.

Section B The President of the Council shall preside at the meetings of the Board of Directors and perform such other duties as required in the bylaws.

Section C The President shall preside at the Annual Business Meeting and meetings of the Executive Committee and be responsible for executive actions in carrying out the policy designated by the Board of Directors. The President may delegate responsibilities within such limits as may be set by the Board to other Officers or Committees.

Section D The President Elect shall preside at meetings of the Board of Directors and of the Executive Committee in the absence of the President. The Senior Vice President and the Vice President shall perform such other duties as may be assigned by the President or the Executive Committee.

Section E The Treasurer shall be responsible for all financial records, collections, and disbursements of the Council. These day-to-day functions will be performed by staff in accordance with Board policy and under the direction of the Treasurer.

Section F The Board of Directors shall employ such professional staff as may be necessary and within the framework of the Council's budget to carry out specified programs and perform necessary duties of the Council.

Section G The Executive Committee shall perform all functions specified in these bylaws and shall represent the Council in all executive functions when the Board of Directors is not in session.

Section H Each Committee and Task Force shall be responsible for the field of activity designated by these bylaws or the Board of Directors, and each shall formulate a program consistent with the objectives of the Council or as instructed by the Board of Directors. An annual report shall be required of the Chairperson of each Committee or Task Force.

Section I The Affiliate Council Liaison shall be responsible for providing input and serving as a consistent contact between the Affiliate Councils and AFGC Board of Directors/Executive Committee on AFGC concerns and interests.

ARTICLE VII
Meetings

Section A There shall be one official business meeting of the Council per fiscal year and such other meetings of the members as determined by the Board of Directors. The business meetings shall include a current financial report and other items approved by the Executive Committee.

Section B The date, time, and place of the Annual Business Meeting or other meetings shall be set by the President, and notices of such meetings sent to all members in good standing at least sixty (60) days in advance of the meeting.

Section C The date, time, and place of the meetings of the Board of Directors shall be set by the President.

Section D The date, time, and place of the meetings of the Executive Committee shall be set by the President.

ARTICLE VIII
Quorum

A quorum shall consist, in the case of the Annual Business meeting and other meetings of members, of 1/40 of the membership in good standing, either in person or by valid proxies (members of Affiliate Councils have only one vote per person, but for quorum purposes only, represent by proxy the entire membership of their Affiliate Council), and in the case of meetings of the Board of Directors and Executive Committee, ½ of the membership of such Board or Committee.

ARTICLE IX
Voting

Section A At the Annual Business Meeting or other meetings of members, actions on items other than national policy issues shall be authorized by a majority of those present in person or by valid proxy.

Section B At meetings of the Board of Directors actions on items other than national policy issues shall be authorized by a majority of those present that are entitled to vote. At meetings of the Executive Committee, actions shall be authorized by a majority of those present who are entitled to vote. In case of dual representation by the same person (example: Board member is also Affiliate Council Advisory Committee President), each voting member only gets one vote.

Section C At meetings of Committees and/or Task Forces actions shall be authorized by a majority of those present.

Section D Actions on national policy issues shall be authorized by a majority of the votes. Actions on national policy issues may only be brought to a vote by the general membership if approved by a 2/3 majority vote of the Board of Directors. If immediate action is required on a national policy issue as determined by the Executive Committee, this action shall be authorized by a 2/3 majority vote of the Board of Directors and affirmed by a 2/3 majority vote at the next annual meeting.

ARTICLE X
Amendments

The bylaws of this Council may be amended by the Board of Directors at any meeting of the Board provided forty-five (45) days prior notice of their consideration is given to the membership.

ARTICLE XI

Rules contained in Robert's Rules of Order shall govern this organization in all cases where they are applicable and are not inconsistent with the bylaws of this Council. A Parliamentarian may be appointed at the discretion of the President.

ARTICLE XII

Directors may participate in a meeting of the Board of Directors by means of telephone conference or similar communications equipment as long as all persons participating in the meeting can hear each other simultaneously.

Adopted 10/2016

PLASMA AMINO ACID CONCENTRATIONS OF HORSES GRAZING ALFALFA, COOL-SEASON PERENNIAL GRASSES, AND TEFF

Amanda S. Reiter, Michelle L. DeBoer, Marcia R. Hathaway, Kerry J. Kuhle, Craig C. Sheaffer, M. Scott Wells, Krishona L. Martinson[1]

The relationship between forage amino acid (AA) profiles, availability of forage AA and plasma AA concentrations of grazing horses is unknown. The objectives of this study were to determine the impact of alfalfa, cool-season perennial grasses (CSG), and teff on plasma AA concentrations of grazing horses. Research was conducted in July 2016 at St. Paul, MN. Alfalfa. Mixed orchardgrass and Kentucky bluegrass (CSG), and teff pastures were grazed by six adult horses who were randomly assigned to one of the three forage types in a Latin-square design. Horses were fasted overnight (12 h) with blood samples and corresponding forage samples taken prior to pasture access (0 h) and after 2 and 4 h of grazing. In general, forage AA concentrations were lowest in teff and highest in CSG, except for tryptophan where no differences were observed. Alfalfa had higher AA concentrations when compared to teff for histidine, isoleucine, lysine, threonine, and valine. Similarly, crude protein values were lowest in teff compared to alfalfa and CSG while acid and neutral detergent fiber values were consistently highest in teff and lowest in CSG. Despite differences in forage AA, threonine was the only plasma AA which differed with higher values observed in horses grazing CSG compared to teff. These results suggest the differences in AA concentrations and nutritive values in the forage may not be strong enough to elicit physiological changes in the horse over short periods of time.

[1]Graduate Research Assistants and Professor, Department of Animal Science, (reit0162@umn.edu; schu2343@umn.edu; hathaway@umn.edu); Veterinarian, Department of Veterinary Population Medicine, (kuhl0004@umn.edu); Professor and Associate Professor, Department of Agronomy and Plant Genetics, (sheaf001@umn.edu; mswells@umn.edu); Associate Professor, Department of Animal Science, (martinson@umn.edu) University of Minnesota, St. Paul, MN, USA

BERMUGAGRASS GERMPLASM TOLERANCE TO BERMUDAGRASS STEM MAGGOT (BSM)

William F. Anderson, Dennis Hancock, Lisa Baxter, Will Hudson [1]

Bermudagrass is the most important perennial forage crop grown for pasture and hay in the Southeast (~ 20 million acres). Bermudagrass Stem Maggot (BSM) (*Atherigona reversura* Villeneuve) is an invasive insect pest that reduces yield of forage bermudagrass hay if it is not controlled. The pest is native to SE Asia, was first documented in North America near Los Angeles, California in 2009 and has since spread throughout the southeast, infesting bermudagrass hayfields as far north as Kentucky and as far west as Texas. A core collection of over 300 forage bermudagrass accessions were evaluated in the field for susceptibility to BSM. Tolerant lines were then evaluated in a replicated field study for yield loss due to the insect by comparing sprayed controlled plots to unsprayed plots in Tifton starting in 2016 through the summer of 2017. For clippings in mid to late summer, BSM reduced yield of Alicia and Russell by over 40% and Tifton 85 by up to 35%. However, some plant introductions in the germplasm collection exhibited less than 10% yield loss and had dry matter hay yields comparable to Tifton 85. These accessions will be further evaluated and used in plant breeding.

[1]Research Geneticist, USDA-ARS, Tifton, GA (bill.anderson@ars.usda.gov); Forage Extension Specialist, Crop and Soils Science Dept., University of Georgia, Athens, GA, (dhancock@uga.edu); Post-doctorate, Crops and Soils Science Dept., University of Georgia, Tifton, GA (baxterl@uga.edu); Extension Specialist, Entomology Dept., University of Georgia, Athens, GA (wghudson@uga.edu).

MANAGEMENT CONSIDERATIONS AND METHODS FOR REPLACING TOXIC TALL FESCUE WITH NONTOXIC NOVEL ENDOPHYTE-INFECTED TALL FESCUE

John Andrae[1]

Tall fescue toxicosis costs the U.S. beef industry over $1 billion annually due to decreased conception rates and gains. These losses can be completely offset by converting pastures from the wild-type endophyte to new cultivars containing a nontoxic endophyte. Even though the economic benefits of conversion are apparent, adoption of this technology has been relatively slow. Real or perceived barriers to converting toxic pastures include 1) complexity of planning a conversion within an existing grazing system, 2) long time period needed for complete replacement, 3) risk of unsuccessful stand establishment, 4) lost forage productivity and/or decreased stocking rates during conversion, 5) risk of recontamination with toxic fescue, 6) risk of losing converted rented land, 7) preference for toxin 'dilution' using legumes, seedhead suppression etc., and 8) issues or concerns with cost and/or cash flow. Many of these issues can be addressed through education and assistance during the conversion planning process. While all grazing operations are unique, prioritizing pastures for conversion based on productivity and animal use is a logical initial step. High priority pastures should maximize economic impacts of smother crops and novel endophytes and encourage additional plantings. Choosing the proper conversion method (i.e. spray-smother-spray or spray-wait-spray) and smother crop will be determined by location, timing, and producer preferences. Good stand management in the year following conversion will also impact novel fescue performance and future acreage conversion.

[1]Professor, Clemson University, Clemson, SC (jandrae@clemson.edu).

IMPACT OF PRENATAL TALL FESCUE TOXICOSIS ON EWE MILK PRODUCTION AND POSTNATAL LAMB GROWTH

J.G. Andrae, J.L. Britt, M.F. Miller Jr., B.M. Koch, S. Adams, J. Klotz, G. Aiken, S.K. Duckett[1]

Tall fescue is the predominant cultivated cool season forage in the United States occupying over 35 million acres for pasture and hay production. Unfortunately, the presence of an endophyte that produces toxic ergot alkaloids significantly reduces animal gains and reproduction. Recently, we conducted two experiments to examine the impact of toxic tall fescue seed fed to ewes during mid and/or late gestation on placental function, fetal growth, milk production, and postnatal muscle growth. In expt. 1, ewes were fed toxic tall fescue seed during mid and/or late gestation and terminal surgeries were performed on d 133 to assess fetal development. Feeding toxic seed during late gestation (d86-133) reduced placentome weights, contractile response in umbilical artery, fetal body weight, and muscle fiber development. Fetal body weight was reduced by 10% in ewes fed toxic tall fescue during late gestation (d86-133). In Expt. 2, we repeated the experiment but the ewes went to parturition for measurement of ewe milk production, postnatal lamb growth and carcass quality. Feeding toxic seed during late gestation (d86-parturition) reduced gestation length by 1.4 d and lamb birth weight by 10%. Ewe milk production in ewes fed toxic tall fescue seed during late gestation was reduced by 57% on d 2 post lambing. Lambs are currently being slaughtered as they reach constant terminal endpoints (final approx. December 10, 2017) and blood samples are being analyzed for age at puberty. These variables and post weaning feed efficiency will be reported at the 2018 AFGC meeting.

[1]Professor, Clemson University, Clemson, SC (jandrae@clemson.edu); Graduate Students, Clemson University, Clemson, SC; Research Scientists, USDA-ARS, Lexington, KY (James.Klotz@ars.usda.gov, Glen.Aiken@ars.usda.gov); Professor, Clemson University, Clemson, SC (sducket@clemson.edu).

STOCKPILING PERENNIAL WARM-SEASON GRASSES IN MISSISSIPPI

B. Rushing[1], R. Lemus[2], M. Thornton[3], and J. Lyles[4]

Stockpiling forage crops enables livestock producers to accumulate feedstuffs into the fall and winter, thus reducing stored forage needs while increasing forage utilization and grazing efficiency. A trial was conducted at Newton, MS in fall, 2015-2017 in which 'Argentine' and 'Pensacola' bahiagrass (*Paspalum notatum*) and 'Common' and 'Cheyenne II' bermudagrass (*Cynodon dactylon*) were evaluated based on their ability to generate nutritious feedstuffs following stockpiling. Four nitrogen (N) rates were applied to each variety: 0, 25, 50, and 75 lb N/acre. Four harvest dates were applied to each N rate at 30, 60, 90, and 120 days after N application. Dry matter (DM) yield was greatly affected by environment. Mean 2015 yield was 1,231 lb DM/acre across all varieties, harvest dates, and N treatments. In 2016, however, mean DM yield was 408 lb/acre across all treatment combinations due mostly to severe droughty conditions. The two bermudagrass varieties out-yielded the two bahiagrass varieties in 2015 and 2016. The 75 lb N/acre rate and the 90 day harvest generated the greatest mean DM yields across all varieties in 2015 and 2016. In terms of forage quality, 'Pensacola' had the highest mean crude protein (CP) and total digestible nutrients (TDN) for both years, with means of 14.0 and 50.2 across all harvests and N rates. Based on data collected thus far, stockpiling warm-season perennial grasses has the potential to generate ample amounts of quality forage for fall and early winter grazing. However, supplementation will be necessary to meet beef cow energy demands.

[1]Assistant Research/Extension Professor; Mississippi State University; Coastal Plain Branch Experiment Station; brett.rushing@msstate.edu; [2]Forage Extension Specialist; Mississippi State University; Department of Plant and Soil Sciences; rlemus@pss.msstate.edu; [3]Forest County Extension Agent; Mississippi State University; mtt52@msstate.edu; [4]Research Associate I; Mississippi State University; Coastal Plain Branch Experiment Station; jcl411@msstate.edu.

EFFECTS OF N APPLICATION AND HARVEST FREQUENCY ON NATIVE WARM-SEASON GRASSES

B. Rushing[1], J. White[2], M. Thornton[3], and J. Lyles[4]

Nine native warm-season grass species combinations were grown in east-central Mississippi on a coarse-loamy, siliceous, semiactive, thermic Glossic Fraguidult, to determine their potential as forage grasses in response to harvest frequency and N application. Established stands were harvested at 30, 40, 60, and 120 d intervals, and were fertilized with N rates of 50, 100, 150, and 200 lb N/a for two consecutive years (2016 and 2017). Year, treated as a repeated measure, was the main source of variation affecting forage yields. Within year, harvest frequency had the greatest effect on cumulative yields in 2016, accounting for 44% of the variation observed. This effect was not observed in 2017. Frequent defoliations combined with droughty conditions from the previous fall led to overall lower yields in 2017. Nitrogen was significant, but moderately, accounting for 2 and 4% of variation for 2016 and 2017, respectively. Species was also significant for both years. Eastern gamagrass [*Tripsacum dactyloides* (L.) L.] was the only species in which cumulative DM yield remained consistent across both years of the study, regardless of harvest frequency or N application. Results suggest that rest intervals should be based on species present, and that less frequent harvests are required for sustained DM yields.

[1]Assistant Research/Extension Professor; Mississippi State University; Coastal Plain Branch Experiment Station; brett.rushing@msstate.edu; [2]Forage Variety Trial Manager; Mississippi State University; Department of Plant and Soil Sciences; jwhite@pss.msstate.edu; [3]Forest County Extension Agent; Mississippi State University; mtt52@msstate.edu; [4]Research Associate I; Mississippi State University; Coastal Plain Branch Experiment Station; jcl411@msstate.edu.

VARIETY MATTERS: WHITE CLOVER VARIETY RESPONSE TO WEED MANAGEMENT PROGRAMS

Byron Sleugh, Pat Burch, Scott Flynn, Ian Corr, and Jeff Nelson

Forage producers face difficult weed management decisions before, during and after establishing forage stands, especially grass-legume mixes. The potential impact on grass-legume forage stands is a concern when weed management plans include the use of herbicides because of the potential negative impact on legumes. This is the single most cited reason by producers for not using herbicides to manage weeds regardless of the species or density of the weeds present. A common question from producers related to the use of herbicides in grass-legume mixtures is "What product do you have that will control weeds and not hurt my clover?" While there are herbicides that show some selectivity (short duration or low levels of injury) on different species of legumes, they often do not meet the farmer's expectation of satisfactory control of a diverse population of problem weeds in pastures. GF- 3731 is an innovative product being developed by Dow AgroSciences that offers the potential of delivering broadleaf weed control while preserving white clover in pastures. Trials were established that evaluated Patriot (Pennington Seed), Durana (Pennington Seed), Barblanca (Barenbrug USA), Alice (Barenbrug USA), and Dutch White Clover (common). We studied the response of these varieties to different herbicides, application timing, and mowing. The results show that the negative impact can be significantly reduced depending on the white clover variety and the choice of herbicide mainly due to less injury and quicker recovery after spraying. These findings indicated that producers are closer to having another tool in an integrated weed management system that will help them meet an important forage management objective – control broadleaf weeds with little or no negative impact on desirable forage plants.

EVALUATION OF PLANTING DATES FOR FALL FORAGE YIELD OF SEVEN WINTER ANNUAL FORAGES

J. Caraway, K. J. Simon, J. A. Jennings, and P. Beck[1]

Livestock producers often plant winter annual forages intended for fall and winter grazing, but are reluctant to plant before October due to the fear of stand failure caused by hot, dry weather conditions. However, delaying planting until October also delays grazeable forage production until late February or early March. Seven winter annual forage species were planted in late August and early October at the University of Arkansas-Southwest Research and Extension Center in Hope to evaluate the influence of planting date on fall dry matter production. The winter annual forage species included: Elbon rye, VNS winter wheat, Jerry spring oat, Coker 227 winter oat, Fridge triticale, WinterHawk annual ryegrass, and Winfred forage brassica. Plots were no-till planted onto a fallowed site. Prior to planting, weeds were sprayed and mowed to provide a uniform stubble height. All the plots were harvested in early December. DM yield for all the treatments was low, < 1,000 lbs./ac., regardless of planting date. This is likely due to a combination of hot and dry weather along with a high percentage of weeds for the August planting. Temperatures were very hot during September, 17 days >90°F, with very little rain for 2 months after planting. August planted Jerry oat was the only early treatment that out yielded the October plantings. This is a case where late summer planting produced good stands but little fall yield. However, most treatments would be expected to produce good forage by late winter to reduce hay feeding at the end of winter.

[1]County Extension Agent - Agriculture, University of Arkansas, Cooperative Extension Service, Texarkana, AR (jcaraway@uaex.edu); Animal Science Program Associate – Forages (ksimon@uaex.edu), Professor and Extension Forage Specialist (jjennings@uaex.edu) , and Professor (pbeck@uaex.edu) University of Arkansas Division of Agriculture. Little Rock, Arkansas. 72204

IN VITRO NEUTRAL DETERGENT FIBER DIGESTIBILITY OF NINE FORAGES IN THE HIGHLANDS AND THE LOWLANDS IN COSTA RICA

C.M. Campos-Granados[1], A. Rojas-Bourrillon[1], A. Martínez-Machado[1]

The objective of this study was to determine the effects of the grazing age in the in vitro digestibility of the neutral detergent fiber of forages in the highlands and the lowlands dairy farms in Costa Rica. Nine forages was evaluated during the entire year, considering the weather season and measuring the phenological age using the technique of counting living leaves. Five samples of every forage were taken considering three different grazing age, depending on each forage management and species. The samples obtained in the field, entered in the CINA laboratories, in order to conduct the respective analyzes: dry matter (DM), neutral detergent fiber (NDF), in vitro digestibility of dry matter at 48 hours (IVDMD) and in vitro digestibility of neutral detergent fiber (NDFD). The data was analyzed using an unrestricted random model comparing the ages of grazing within highland forages (*Lolium perenne, Kikuyuocloa clandestine, Festuca* sp. x *Lolium* sp., *Cynodon nlemfuensis*) and lowland forages (*Brachiaria brizantha* var. Mulato II, *Brachiaria mutica, Hemarthria altissima, Panicum maximum* var. Mombaza, *Ischaemum ciliare*) using the Duncan-Waller test. The data show that age of grazing has a direct impact on cell walls of forages and dry matter contents in both highland and lowland forages, and this impacts the energy content of these materials, which translates into direct effects on milk production. In addition, it can be concluded that highland forages showed higher values of NDFD with respect to lowland forages, which partly explains their greater potential for milk production in the tropics.

[1]Profesor. Escuela de Zootecnia y Centro de Investigación en Nutrición Animal (CINA), Facultad de Ciencias Agroalimentarias, Universidad de Costa Rica. San José, Costa Rica. carlosmario.campos@ucr.ac.cr, augusto.rojas@ucr.ac.cr, amartínez@feednet.ucr.ac.cr.

EFFECT OF RYZUP AND AMS ON HAY YIELD IN OHIO

Bruynis, C.L., and McCutcheon, J. S.[1]

Farmers are always looking for opportunities to jump-start the grazing season to reduce the need for stored forages. RyzUp SmartGrass® (Valent) is a growth hormone (gibberellin A) being offered to producers to enhance spring grass growth. Previous research in Ohio has indicated a slight yield response but the results were highly variable. This research was designed to test the effect of adding ammonium sulfate (AMS) and Ryzup on orchardgrass growth in Ohio. The experiment used a randomized block design with four replicated plots. Treatments included a control, AMS, Ryzup, and AMS plus Ryzup applied on March 18, 2017. Growth was determined for each plot by taking 30 measurements of a calibrated rising plate meter 28 days after application. Harvest yield was measured using a plot harvester on May 16, 2017. Data from the plate meter indicated that plots receiving AMS, whether alone or with RyZup, had more growth. These treatments were significantly different ($P < 0.05$) that the plots receiving Ryzup alone or the control plot replications. Cutting yield data showed the Ryzup plus AMS treatment tended different ($P > 0.10$) than the other treatments, yielding approximately 500 to 600 pounds more lbs/acre of forage.

[1]Bruynis, C.L., OSU Extension, Ross County, 475 Western Ave., Suite F, Chillicothe OH 45601 bruynis.1@osu.edu; McCutcheon, J.S., OSU Extension, Southeast Region Office, 16714 Wolf Run Rd (SR 215), Caldwell, OH 43724-9414, mccutcheon.30@osu.edu.

MOWING TO CONTROL WEEDS IN PASTURES

M. A. Landefeld, C. D. Penrose, T. G. Wiseman, J. S. McCutcheon[1]

Weed control is an essential part of all forage production systems because unwanted weeds compete for nutrients, water and sunlight needed for optimum growth. Forage growth in Ohio pastures is a critical part of farm production because grazing livestock are present on more than 36 percent of Ohio farms. Weeds can reduce forage crop quality and productivity if left uncontrolled. The purpose of this study was to determine if weed populations in pastures could be changed or reduced by varying the timing of mechanical mowing throughout the late spring and summer growing period without the use of herbicides. This 32-plot randomized complete block study was conducted in 2015 and 2016 at the Eastern Agriculture Research Station, Belle Valley, OH. Treatments in this study consist of: (1) Control (no mowing), (2) June mowing, (3) July mowing, (4) August mowing, (5) September mowing, (6) June/August mowing, (7) July/September mowing, and (8) June/July/August/September mowing. A scale of 0-9 was used (0 meaning 0% weeds present to a 9, meaning 90% weeds present). Results indicate all of the mowing treatments had significantly less weeds present ($P<0.05$) than the control except for the June only treatment. The two-mowing treatment of June/August and the four-mowing treatment of June/July/August/September were significantly less ($P<0.05$) than both the non-mowing treatments and the June only treatment. This study suggests that the June/August mowing combination may be the best option to reduce weeds.

[1]Ohio State University Educator, Woodsfield, OH (landefeld.6@osu.edu); Ohio State University Extension Educator, McConnelsville, OH (penrose.1@osu.edu); Ohio State University Extension Educator, Somerset, OH (wiseman.15@osu.edu); Ohio State University Extension Southeast Region Director, Caldwell, OH (mccutcheon.30@osu.edu).

STOCKPILING ORCHARDGRASS

M.A. Landefeld, D.F. Lima, J.S. McCutcheon, C.D. Penrose[1]

Abstract

Orchardgrass can be an option to extend the grazing season. The purpose of this study was to determine the effects of yield and quality by adding urea, urea with a fertilizer stabilizing (urease inhibitor) product, and ammonium sulfate by applying them to orchardgrass. There were no statistical differences in CP, ADF, and TDN. There was a statistical ($P < 0.05$) difference in yield with of 3694 lbs DM/acre with the urea plus Agrotain®, compared to the control (2698 lbs DM/acre). In many cases, stockpiling is a viable option to reduce costs and save time, and if dry weather is forecasted when applying urea, utilizing a urease inhibitor may increase yield and quality.

Introduction: Stockpiling cool season grasses for fall and winter grazing is an economical way to reduce stored feed costs. Fescue *(Lolium arundinaceum)* is a common grass that is stockpiled, but orchardgrass *(Dactylis glomerata)* is also a common grass in many parts of the country. There is limited data on stockpiling orchardgrass. However, research by Riesterer, Undersander, Casler, and Combs (2000), Pope (2011), and Penrose, Landefeld and McCutcheon (2017, submitted) indicates that stockpiling orchardgrass can be an option. When stockpiling, nitrogen is often used to increase yield and quality. Grass plants use nitrogen to maximize growth, produce proteins, and build-up sugars for growth. Livestock are then permitted to graze the "stockpile" at a later date when other forages no longer are growing or available (Landefeld & Penrose, 2016). Ammonium sulfate and urea are two forms of nitrogen commonly available. Urea is more frequently used due to the availability of product, the cost compared to ammonium sulfate, and equipment available for making applications. One problem utilizing urea with this practice is the possibility of losing broadcast nitrogen due to volatilization because of inadequate rainfall in a timely manner. Volatilization losses may be reduced if a urease inhibitor is used with the urea. The most common urease inhibitor is NBPT (N-(n-butyl) thiophosphoric triamide) sold under the trade name Agrotain® (Agrotain International) (Schwab, & Murdock, 2010). In a previous study Penrose, McCutcheon & Landefeld (2015) concluded, stockpiling cool season grasses, and in particular fescue can reduce the need for stored forages and in many cases it is higher quality than the hay producers make. In that study, there was a positive statistical difference ($P < 0.05$) in the amount of crude protein above the control plots when adding Agrotain® to the urea before application. However, there is limited data on the impact of using a urease inhibitor on stockpiling orchardgrass. The purpose of this study was to determine the effects of yield and quality by adding urea, urea with a fertilizer stabilizing (urease inhibitor) product, at the labeled rate, and ammonium sulfate by applying them to orchardgrass.

[1]Assistant Professor and Educator, Ohio State University Extension; Extension Educator, Ohio State University Extension; Associate Professor and Regional Director, Ohio State University Extension; Associate Professor and Educator, Ohio State University Extension, respectively.

Materials and Methods: This was a randomized complete block design with four (4) treatments, including a control, and four (4) replications of each treatment. Each plot was six feet by 20 feet. The site was a predominately orchardgrass hay field, and the soil test results were pH, 6.2; P, 26 ppm; K, 68 ppm; Ca, 969 ppm; and Mg, 209 ppm. The field was mechanically harvested to a height of 4 inches prior to the treatment. The control plots received no ammonium sulfate, urea or urease inhibitor. For the other treatments, urea nitrogen was broadcast on the surface as follows: 100 lbs. urea (46-0-0)/acre, 100 lbs. urea/acre plus Agrotain® added at the labeled rate of one gallon*/per ton of fertilizer, and 219 lbs ammonium sulfate (21-0-0)/acre. The application date was August 2, 2016. The plots were harvested on November 4, 2016 to a height of three inches above ground level utilizing 2' x 2' samples from each plot. Each sample was weighed fresh, and then taken to a laboratory for forage analysis. Each of the 16 samples were quality tested for Crude Protein (CP), Acid Detergent Fiber (ADF) and Total Digestible Nutrients (TDN).

Statistics were calculated using Proc Mixed in SAS 9.3. The model included treatment, and the random variable of replication. Pdiff was used for mean separation compared to the control.

Results and Discussion: In the 2016 study (Table 1), there were no statistical differences in CP, ADF, and TDN. There was a statistical ($P < 0.05$) difference in yield with of 3694 lbs DM/acre with the urea plus Agrotain®, compared to the control (2698 lbs DM/acre). The treatment of urea alone (3119 lbs DM/acre) and ammonium sulfate (3372 lbs DM/acre) were not statistically different ($P > 0.05$) than the control.

In a 2015 study, Landefeld and Penrose (2016) reported no statistical difference in DM production by adding urea, urea with Agrotain®, or urea with Nutrisphere®. This may be attributed to lack of rainfall in the first 30 days of that study, only totaling 1.10 inches. However, they did report an increase in CP in the urea with Agrotain® treatment. Initiation dates and the amount of nitrogen applied were similar for both studies. The date of final harvest was one month earlier (December 2, 2015 vs. November 4, 2016) in this trial because (Henning, Wheaton & Roberts, 1993) noted that orchardgrass tends to lose quality faster than fescue.

Stockpiling initiation date can have an effect on the amount of forage that accumulates. When stockpiling begins earlier in the season (i.e. July 1 vs. August 1), we would expect to see more tonnage, but lower quality in stockpiled forage started in July than that in August, if harvested at the same date. Amount of rainfall can also have tremendous effects on stockpiling quantity and quality. Rainfall in the first 30 days from trial initiation in 2016 was 3.2 inches, but the first significant rainfall (0.7 inches) was twelve days after initiation and daytime high temperatures averaged 89.6°F the first two weeks of the trial. Researchers utilized the weather forecast to maximize the effects of using a urease inhibitor by starting the trial when little or no rain was forecast. This could explain the difference in yield with the urea plus Agrotain®.

In this study, CP averaged 12.28% for the control, 12.40% for the urea only, and 12.65% for the urea plus Agrotain®. While there were expected differences in the CP, they were not significant ($P < 0.05$). CP from the ammonium sulfate treated plots was the greatest numerically of the treatments, but not significantly different. The yield was numerically less than the plots treated with urea and urea plus Agrotain®. Adding urea (100 lbs/acre) to the stockpiled forage at a rate of 46 lbs N/acre cost $19/acre when urea is $385/ton. Including Agrotain® at the labeled rate adds $4 per acre for a total of $23/acre. When ammonium sulfate (219 lbs/acre) at 46 lbs

N/acre per acre is applied, the cost is $88/acre for the fertilizer. These amounts do not include application costs that would be incurred.

There are several factors to consider prior to stockpiling including cost of the fertilizer, application costs, the costs and time to feed stored forages and grain, and the potential success and utilization of the stockpiled forages. There are still variables in the success of stockpiling such as rainfall, nitrogen availability, and temperatures while stockpiling, and winter weather conditions when grazing the orchardgrass. In many cases, stockpiling is a viable option to reduce costs and save time. If dry weather is forecasted when applying urea, utilizing a urease inhibitor may increase yield and quality.

Literature Cited:

Henning, J., Wheaton, H, & Roberts, C. (1993). Tall Fescue. University of Missouri Extension Fact Sheet, G4646. Retreived from: http://extension.missouri.edu/p/G4646

Landefeld, M. & Penrose, C. (2016). Comparing Fertilizer Additives Applied to Urea for Stockpiling Orchardgrass. Available at: http://www.nacaa.com/journal/post_editor.php?article_id=549

Penrose, C., McCutcheon, J. & Landefeld, M. (2015). Utilizing Urease Inhibitors with Urea to Stockpile Fescue. Available at: http://www.nacaa.com/journal/index.php?jid=549

Penrose, C., Landefeld, M. & McCutcheon, J. (2017, submitted). Fertilizing Options for Stockpiled Forages. https://agcrops.osu.edu/on-farm-research

Paine., L., & Bennett, K. (2009), Stockpiling Pasture, Available at: http://www.midwestforage.org/pdf/403.pdf.pdf

Pope, J. (2011). Stockpiling Forage Grasses for Winter Grazing in West Tennessee. Retreived from: https://www.utm.edu/departments/msanr/_pdfs/Pope%20Research%20Project%20Final.pdf

Riesterer, Undersander, Casler, and Combs (2000). Forage yield of stockpiled perennial grasses in the upper midwest USA. Agronomy Journal. 92:740-747.

Schwab, G., & Murdock, L. (2010). Nitrogen Transformation Inhibitors and Controlled Release Urea. Retrieved from: http://www2.ca.uky.edu/agc/pubs/agr/agr185/agr185.pdf

Table 1. Yield and quality of stockpiled orchardgrass

Treatments	lbs DM/acre	lbs DM/acre above control	CP%	ADF%	TDN%
Control	2698	-	12.28	41.31	61.60
Urea	3119	421	12.40	42.80	60.56
Urea+Agrotain®	3694*	996*	12.65	41.83	61.24
Ammonium Sulfate	3372	674	12.94	43.43	60.13

LSD = 981 ($P < 0.05$) * denotes significant difference in yield.

RISK AND RETURN FROM GRAZING WARM SEASON GRASSES

C.N. Boyer, K. Zechiel, J. Rhinehart, P. Keyser, and G. Bates[1]

Tall fescue (TF) is a cool-season grass that is adaptable, easy to establish, and persistent under adverse conditions, which is why cattle producers primarily rely on it for pasture and hay in the Fescue Belt. However, physiological characteristics of TF can cause problems for cattle producers during the summer. During summer, cattle grazing endophyte-infected TF are likely impacted by fescue toxicity, which can result in elevated body temperature, lower conception rates, reduced average daily gain, and failure to shed winter coat, resulting in losses of over $1 billion annually. A possible solution to this problem is to rotate cattle to warm-season grasses (WSGs) during summer when impacts of toxicosis are typically greatest. Thus, the objective of this study was to analyze the profitability and risk associated with grazing five different WSGs in Tennessee. Data comes from a 3-year (2014-2016) experiment in two Tennessee locations. WSGs included switchgrass (SG), bermudagrass (BG), eastern gamagrass (EG), big bluestem/indiangrass (BBI), and crabgrass (CG). Partial budgeting was used to estimate net returns to grazing and a simulation analysis was performed to assess risk exposure. Grazing perennial WSGs (SG, BBIG, EG, and BG) were more profitable than grazing the annual (CG). SG had the highest expected net returns and lowest risk exposure than all the other WSGs. Economic results will help producers make profitable grazing decisions during the summer months.

[1] Associate Professor, Department of Agricultural and Resource Economics at The University of Tennessee (UTK), Knoxville, TN (cboyer3@utk.edu); Graduate Student, Department of Animal Science at UTK, Knoxville, TN (kzechiel@utk.edu); Associate Professor, Department of Animal Science at UTK, Knoxville, TN (jrhinehart@utk.edu); Professor, Department of Forestry, Wildlife, and Fisheries at UTK, Knoxville (pkeyser@utk.edu), TN; Professor, Department of Plant Science at UTK, Knoxville, TN (gbates@utk.edu).

COW/CALF MANAGED INTENSIVE GRAZING FROM SCRATCH – A LEARNING CURVE

Christopher R. Gibbs[1]

Purchasing highly erodible and environmentally sensitive land can be difficult, if not impossible, to cash-flow in western Ohio without the introduction of livestock. This real time, real life, real people, and real dollars trial sought to determine if marginal land could be sustainable, both environmentally and economically, by introducing cattle and managing them using an intensive grazing system. In 2010, Gibbs Farms LLC purchased a 226 acre tract of primarily highly erodible land containing 25 acres of cropland suitable for annual row crops, 86 acres of woodlands, and 115 acres of contiguous abandoned pastureland. Beginning in the fall of 2010, numerous management initiatives and/or land treatments were introduced to the whole tract including contracting for custom grazed cattle, clearing fence lines, building high tensile electrified line and internal paddock fencing, removing black locust and brush, installing sub-surface water line, installing geo-thermal frost-free livestock waters, drilling and installing new wells with sub-frost line pressure mechanisms and controls, frost seeding red and white clovers, planting cover crops on cropland to graze or hay prior to spring row cropping, building cattle and manure housing, and executing a Managed Intensive Grazing system. The results of this trial indicate that the environmental sustainability of the land has been maintained and improved. The economic sustainability of the operation is meeting cash flow needs and increasing net worth to the Limited Liability Company; however, returns to labor and management are more elusive.

[1]Owner/ Rancher Maplewood Ohio, cgibbs58@gmail.com.

THE EFFECTS OF GIBERELLINS AND UREA ON SPRING FORAGE GROWTH

D. F. Lima, C. H. Gelley, M. A. Landefeld, J. S. McCutcheon[1]

Gibberellins are compounds classified as plant growth regulators. This class of compounds has been shown to regulate multiple plant developmental factors such as breaking seed dormancy, and stem and leaf elongation. Our experimental study focused on the latter to increase spring forage growth utilizing RyzUp SmartGrass®, a gibberellin based product. Previous studies have demonstrated that gibberellins have increased biomass in mixed forages before, but we wanted to test it against a typical fertilizer rate application. This experiment was designed to assess the benefits of utilizing a hormone treatment for plant growth with and without a traditional fertilizer. Four treatments were implemented; a control with no added Urea or RyzUp, Urea applied at 100 lb./ac., RyzUp applied at 4 oz./ac., and both RyzUp and Urea applied together in the early spring. Our group hypothesized that we would obtain an increase in biomass utilizing RyzUp SmartGrass® however, the rate would not be as great as the nitrogen application in the form of urea. Results of the 2017 trial showed no significant differences ($P<0.05$) in the control and experimental groups except for an approximate 10% increase in biomass production when the gibberellin growth regulator was co-applied with urea. Possible reasons for no response to urea alone could be unseasonably warm temperatures volatilizing the urea post application. Explanation for the response treatment is that there was a synergistic effect when urea was co-applied with the gibberellin, further investigation will be pursued.

[1]Ohio State University Extension Educator, St. Clairsville, OH (lima.19@osu.edu); Ohio State University Extension Educator, Caldwell, OH (gelley.2@osu.edu); Ohio State University Educator, Woodsfield, OH (landefeld.6@osu.edu); Ohio State University Extension Southeast Region Director, Caldwell, OH (mccutcheon.30@osu.edu).

QUANTIFYING MORPHOLOGICAL DEVELOPMENT AND FORAGE NUTRITIVE VALUE OF STEM AND LEAF FRACTIONS FOR REDUCED LIGNIN ALFALFA

A.M. Grev[1], M.S. Wells[2], D.N. Catalano[1], K.L. Martinson[3], and C.C. Sheaffer[4]

This study was conducted to characterize changes in morphological development and forage nutritive value of alfalfa stem and leaf fractions for the reduced lignin alfalfa variety '54HVX41' and the reference alfalfa variety 'DKA43-22RR'. Alfalfa was seeded in a randomized complete block at two locations in Minnesota and hand-sampled at weekly intervals for a period of five to six weeks during the summer of 2015 and the spring and summer of 2016. Samples were staged to determine maturity and then divided into stem and leaf fractions, weighed, and analyzed for forage nutritive value. On average, sampling began around the mid-vegetative stage and finished after flowering. Forage maturity, leaf:stem, stem and leaf crude protein (CP), stem and leaf neutral detergent fiber (NDF), leaf NDF digestibility (NDFD), and leaf acid detergent lignin (ADL) were similar for both alfalfa varieties. Stem ADL concentrations increased across time ($r^2 = 0.78$) for both varieties but were lower for '54HVX41' (average 5.4% ADL) compared to 'DKA43-22RR' (average 6.3% ADL). Stem NDFD decreased across time ($r^2 = 0.91$) for both varieties but was greater for '54HVX41' (average 40% NDFD) compared to 'DKA43-22RR' (average 36% NDFD). These results indicate that much of the change in forage nutritive value for reduced lignin alfalfa is a result of changes within the stem fraction of the plant, and that these changes are present across a range of forage maturities.

[1]Graduate Research Assistant, Department of Animal Science, University of Minnesota (grevx004@umn.edu; catal039@umn.edu); [2]Assistant Professor, Department of Agronomy and Plant Genetics, University of Minnesota (mswells@umn.edu); [3]Associate Professor, Department of Animal Science, University of Minnesota (krishona@umn.edu); [4]Professor, Department of Agronomy and Plant Genetics, University of Minnesota (sheaf001@umn.edu)

ASSESSING MINERAL CONCENTRATIONS OF COOL-SEASON GRASSES ACROSS SEASONS

A.L. Fowler, C. Teutsch, R. Smith, B. Goff, K. Prince, and L.M. Lawrence[1]

Pasture mineral concentrations may vary during the growing season, making it difficult to precisely formulate supplementation programs for grazing animals. Near-infrared spectroscopy (NIRS) has the potential to provide rapid information about pasture composition, but has not been widely used for mineral analysis. This study developed NIRS equations to predict Ca, P, and Mg concentrations in four cool-season grasses (bluegrass, orchardgrass, perennial ryegrass and tall fescue) sampled from April to November. Half of the plots received N in March, May, and August. Wet chemistry was performed on a sub-set of samples to develop prediction equations with WinISI® software (v.4.4) for Ca, P, and Mg after all samples were scanned (FOSS 6500 NIRS). The R^2 for all equations was ≥ 0.90 and the SE of cross-validation was ≤ 0.06. The concentrations of Ca, P and Mg fluctuated throughout the growing season and were affected by species. The following ranges were observed: P, 0.391 to 0.678%; Ca, 0.283 to 0.758% and Mg, 0.113 to 0.347%. The Ca:P also varied (range 0.50:1 to 1.58:1). Nitrogen increased P and Mg concentrations, but had no effect on Ca. Rapid NIRS determination of pasture minerals could be used to develop more precise and dynamic dietary supplementation programs for grazing animals.

[1]Department of Animal and Food Sciences, University of Kentucky, Lexington, KY (ashleyfowler@uky.edu); Department of Plant and Soil Sciences, University of Kentucky, Lexington, KY (chris.teustch@uky.edu); Department of Plant and Soil Sciences, University of Kentucky, Lexington, KY (raysmith1@uky.edu); Department of Plant and Soil Sciences, University of Kentucky, Lexington, KY (ben.goff@uky.edu); Department of Plant and Soil Sciences, University of Kentucky, Lexington, KY (kelly.prince@uky.edu); Department of Animal and Food Sciences, University of Kentucky, Lexington, KY (llawrenc@uky.edu)

HEAT TOLERANT ANNUAL RYEGRASS: ALLOWING FOR EARLIER ESTABLISHMENT

E.D. Billman, J.I. Morrison, and B.S. Baldwin[1]

Annual ryegrass (*Lolium multiflorum*) is a cool-season forage used as a feedstuff for beef operations throughout much of the southeastern United States. Late summer or early fall planting of annual ryegrass is of high desirability, as it increases available forage to cattle earlier in the season and allows for increased seasonal weight gain. However, temperatures during late summer and early fall often exceed 90 °F in southern states that rely on annual ryegrass for winter grazing, such as AL, GA, FL, LA, MS, and TX. The objective of this research was to develop heat tolerant annual ryegrass through recurrent phenotypic selection that will germinate and survive when planted in high temperature conditions earlier in the growing season. Seed of the cultivar Marshall and subsequent cycles of selection were subjected to high temperature germination screening in growth chambers at 104 °F in 2015, 2016, and 2017. Seedlings that successfully germinated and survived for 14 d advanced to cycle 1, cycle 2, and cycle 3, sequentially. In 2017, a significant increase in germination at 104 °F was observed between cycle 2 (18%) and cycle 0 (3%). A significant delay in days to maturity was also observed between cycle 2 (115 d) and cycle 0 (105 d). The results indicate that selection for quantitative traits in annual ryegrass is possible with recurrent phenotypic selection. Selection for heat tolerance concurrently affected maturity, indicating the two traits may be linked.

[1]Ph.D. Student, Dept. of Plant and Soil Sciences, Mississippi State University (edb257@msstate.edu); Research Associate III, Dept. of Plant and Soil Sciences, Mississippi State University (jim46@pss.msstate.edu); Professor, Dept. of Plant and Soil Sciences, Mississippi State University (BBaldwin@pss.msstate.edu).

THE FORAGE YIELD AND NUTRITIVE VALUE OF HEMP AS DETERMINED BY PLANTING DATES AND PHENOLOGICAL STAGE

C.E. Stringer, B.M. Goff, R.C. Pearce, D. Hildebrand, B.S. Baldwin[1]

In the years following the 2014 Farm Bill, hemp has become a tremendously popular crop in U.S. states permitting its cultivation and research. Much interest has been given to furthering research about this crop, assessing its suitability for this region, and evaluating it for a variety of consumer and industrial applications. This study was conducted to determine the overall potential of hemp as a forage and how its forage quality changes over the course of the growing season. Three varieties of hemp (a grain, a fiber, and a dual purpose type) and one variety of kenaf were evaluated over two planting dates (a "typical" and a "late"). Samples were taken every two weeks from May until first frost (October). Data was collected for yield, population, phenological stage, morphological composition in the field; samples were then dried and ground for lab analysis, which includes crude protein (CP), digestibility (IVTD), neutral detergent fiber (NDF), acid detergent fiber (ADF), and acid detergent lignin (ADL). Planting date was non-significant in the first year's data. Based on our lab analysis of the 2016 data, kenaf remains a better forage than hemp in terms of yield and maintaining desirable levels of digestibility, lignin, and NDF later in the growing season than hemp. While hemp and kenaf have potential uses in ruminant management systems, their applications currently remain limited to specialty markets or extreme situations such as an alternative use for poor stands of grain and fiber hemp.

---

[1] Graduate student, University of Kentucky, Lexington, KY (cst256@uky.edu); Assistant Professor of Forages, University of Kentucky, Lexington, KY (ben.goff@uky.edu); Extension Professor of Tobacco at University of Kentucky, Lexington, KY (rpearce@uky.edu); Research Professor at University of Kentucky, Lexington, KY (dhild@uky.edu); Professor at Mississippi State University, Starkville, MS (bsb2@msstate.edu)

YIELD, PREFERENCE, AND FORAGE NUTRITIVE VALUE OF WINTERHARDY PERENNIAL RYEGRASS UNDER ANIMAL GRAZING

D.N. Catalano[1], C.C. Sheaffer[2], A.M. Grev[1], N.J. Ehlke[3], K.L. Martinson[4]

Perennial ryegrass (PR) is a high quality and yielding forage that, until recently, has lacked the winterhardiness to persist in the upper Midwest. The objectives were to evaluate new winterhardy perennial ryegrass varieties under livestock grazing. Research was conducted in St. Paul (SP) and Grand Rapids, MN (GR). Meadow fescue (MF), tall fescue (TF), orchardgrass (OG), and three perennial ryegrass varieties ('Forageur', 'Remington', and 'Spreader IV') were seeded in 2015 (SP) and in 2016 (GR); and were grazed for 2 days by either 6 adult horses (SP) or by 6 cow-calf pairs (GR). Prior to grazing, forage yield and quality samples were collected, and post-grazing, fields were visually assessed for the percentage of forage removal to determine livestock preference. Forage yields of PR varieties were similar but lower yielding (2.9 to 3.1 tons/acre in 2016, 1.0 to 1.1 tons/acre in 2017). At both locations, PR varieties were among the most preferred (75 to 94% removal) while OG was the least preferred. PR varieties were consistently among the highest for crude protein (19 to 22% dry matter [DM]), equine digestible energy (1.1 mcal/lb), bovine metabolizable energy (1.1 to 1.2 mcal/lb), nonstructural carbohydrates (11% DM), and among the lowest for neutral detergent fiber (50 to 53% DM). These results confirm that winterhardy varieties of PR are preferred by horses and beef cows and have superior nutrient levels compared to other perennial cool-season grasses. This research will be continued into 2018 to assess long-term winter survival of the new forage varieties.

[1]Research Assistant, Dept. of Animal Science (catal039@umn.edu); Research Assistant, Dept. of Animal Science (grevx004@umn.edu); [2]Professor, Dept. of Agronomy and Plant Genetics, (sheaf001@umn.edu); [3]Professor and Dept. Head, Department of Agronomy and Plant Genetics (ehlke001@umn.edu); [4]Associate Professor, Dept. of Animal Science (krishona@umn.edu); University of Minnesota, St. Paul, MN

ALTERNATIVES TO CONVENTIONAL NITROGEN FERTILIZATION ON TALL FESCUE AND BERMUDAGRASS

M.D. Corbin[1], R.L.G. Nave, G.E. Bates, D.M. Butler and S.A. Hawkins

Alternatives to conventional nitrogen (N) fertilization on tall fescue and bermudagrass were studied at the University of Tennessee Plateau Research and Education Center in Crossville, TN. Experimental period occurred from April-September 2016 and 2017, and the experimental design for each experiment was a completely randomized block design with six treatments and four replications per treatment (n = 24). Experiment 1: tall fescue and experiment 2: bermudagrass. For both experiments treatments were as followed: 1) control (CN) without N fertilization; 2) grass and white clover (WC) at a rate of 4 lb/acre; 3) grass and red clover (RC) at a rate of 8 lb/acre; 4) grass and cowpea (CW) at a rate of 50 lb/acre; 5) fertilization with broiler litter at a rate of 2 ton/acre; and 6) fertilization with ammonium nitrate (AN) at a rate of 60 lb/acre. For each analysis, the dependent variable was herbage mass (HM), crude protein (CP), acid detergent fiber and neutral detergent fiber. Red clover treatments resulted in highest for total HM for 2017 in tall fescue highest for HM for 2016 and 2017 in bermudagrass. Treatments containing BL and AN showed no differences for total HM and CP for 2016 and 2017 in both experiments. Treatments containing CW presented over-all lower results compared to other treatments. Utilizing these results in combination with cost associated with each source could assist producers in choosing a productively efficient source of N or a method of reducing their amount and annual-overall cost associated with conventional N.

[1]Research Associate, Plateau Research and Education Center, University of Tennessee, mcorbin1@utk.edu

EFFECTS OF GRAZING MANAGEMENT AND MONENSIN ON COW-CALF PERFORMANCE AND PASTURE PRODUCTION

Leanna M. Duppstadt*, Sara K. Linneen, Benjamin C. Williamson, Jessica A. Williamson[1]

Monensin (Rumensin 90®, Elanco Animal Health) is a direct-fed ionophore that alters ruminal fermentation and has been shown to increase cow energy efficiency by altering VFA production in the rumen, resulting in improved forage utilization; suggesting that monensin has the potential to alleviate the effects of a cool-season perennial pasture "summer slump". An experiment was conducted using 32 spring-calving, Angus-based cow-calf pairs randomly allocated to pastures based on body weight (BW) and blocked by age. The study was conducted for 3 years (2015-2017) over a 56-d grazing season and set up as a 2x2 factorial with all cows receiving limited nutrition wheat mid-based pellets either with (MON; 200 mg/cow/day) or without monensin (NOMON) on rotational (ROT) or continuously (CONT) grazed pastures. ROT were rotated every 7-d and desired forage height was maintained by put-and-take stocking. Precipitation was above average in June and July for Years 1 and 3, and critically below average in Year 2. There were no significant grazing x supplement interactions or main effects for animal performance for Years 1 and 3 ($P < 0.10$). In Year 2, all cows lost BW and body condition (BCS) from d 0 to 56, but cows on ROT receiving MON lost the least (grazing x supp: BW ($P < 0.01$) and BCS ($P = 0.02$)). There were no significant effects or interactions for carrying capacity (CC) or stocking rate (SR) in Years 1 and 3. ROT MON had significantly greater CC and SR ($P = 0.07$ and $P < 0.01$, respectively) compared to CONT NOMON. MON during times of ample forage availability and quality appears to have no significant impact; however, during times of drought stress, MON may benefit pasture productivity and cow performance.

[1]Master of Science Candidate, The Pennsylvania State University, University Park, PA (lms5900@psu.edu); Animal Science Dept., Penn State (sara.linneen@gmail.com); Animal Science Dept., Penn State (bcw13@psu.edu); Plant Science Dept., Penn State (jaw67@psu.edu).

GLUCOSE AND INSULIN RESPONSE OF HORSES GRAZING ALFALFA, PERENNIAL COOL-SEASON GRASSES AND TEFF ACROSS SEASONS

M.L. DeBoer, M.R. Hathaway, A. Reiter, K.J. Kuhle, P.S.D. Weber, C.C. Sheaffer, M.S. Wells, and K.L. Martinson[1]

Elevated nonstructural carbohydrate (NSC) values in forages can cause adverse health effects in horses diagnosed with metabolic disorders, yet many owners still demand pasture access for these horses. The objectives of this research were to determine the impact of different forage species on blood glucose and insulin concentrations of horses throughout the grazing season. Research was conducted in May (Spring), July (Summer), September (Fall), and October (Late Fall) 2016 in St. Paul, MN. Alfalfa, mixed orchardgrass and Kentucky bluegrass (CSG), and teff grass pastures were grazed by six adult horses who were randomly assigned to one of the three forage types in a Latin-square design. Horses were fasted overnight (12 h) with blood samples and corresponding forage samples taken at 0, 2, 4, 6 and 8 h after grazing. Teff grass consistently had lower digestible energy, crude protein and NSC, and higher neutral and acid detergent fiber. Consequently, horses grazing teff grass had lower peak glucose values in the late fall and lower peak insulin values in both the fall and late fall when compared to horses grazing CSG. These results suggest teff grass may be a pasture forage option for horses diagnosed with metabolic disorders.

[1]Graduate Research Assistant, Professor, and Graduate Research Assistant, Department of Animal Science, University of Minnesota, St. Paul, MN (schu2343@umn.edu; hathaway@umn.edu; reit0162@umn.edu); Veterinarian, Department of Veterinary Population Medicine, University of Minnesota, St. Paul, MN (kuhl0004@umn.edu); Assistant Professor, Department of Large Animal Clinical Sciences, Michigan State University, East Lansing, MI, USA (weberp@cvm.msu.edu); Professor and Associate Professor, Department of Agronomy and Plant Genetics, University of Minnesota, St. Paul, MN, USA (sheaf001@umn.edu; mswells@umn.edu); Associate Professor, Department of Animal Science, University of Minnesota, St. Paul, MN (martinson@umn.edu).

TIME BUDGETS OF MARES AND FOALS GRAZING COOL-SEASON GRASS PASTURES

M.B. Pyles, S.H. Hayes, A.L. Fowler, V.T. Bill, A.D. Crum, and L.M. Lawrence[1]

Nonstructural carbohydrates (NSC) accumulate in cool-season grasses (CSG) under sunny conditions; thus NSC concentration in CSG tends to be lowest in the morning and highest in the evening. We hypothesized that mares and foals would spend more time grazing when NSC would be expected to be highest. Three pastures of Thoroughbred mares (2 mares/pasture) and 2 pastures each with two mare-foal pairs were observed in August, 2017 when foals were 3-5 months of age. Instantaneous scan samples of behavior were recorded every 5 min from 7:00 AM to 7:00 PM. Mares spent 56% of the day consuming forage and 34% of the day standing. Mares spent significantly more time foraging in the sun during midday hours (11:00 AM – 3:00 PM) compared to the morning hours (7:00 AM – 11:00 AM), but there were no differences between morning and evening hours (3:00 PM – 7:00 PM). Time of day did not affect foraging behavior of foals. Foals spent less time than mares consuming forage (38% of the day), but similar amounts of time standing (30%). Foals spent more time lying down compared to mares (13% and 0.4% of their day, respectively). In this study, mares did not spend more time foraging in the evening at the time of day that NSC content of CSG is expected to be the highest.

[1]Graduate Research Assistant, (morgan.brewer@uky.edu); Research Analyst, (shhayes@uky.edu); Graduate Research Assistant, (afo227@uky.edu); Graduate Research Assistant, (vbill14@gmail.com); Senior Laboratory Technician, (andrea.crum1@uky.edu); Professor, (llawrenc@email.uky.edu); all from the Department of Animal and Food Sciences, University of Kentucky, Lexington, KY.

FORAGE QUALITY AND CARBOHYDRATE CONTENT OF COOL SEASON PERENNIAL GRASSES

A.C. Riley, K. Prince, and S.R. Smith[1]

Cool season perennial grasses are the backbone of equine nutrition in Kentucky and across the transition zone. The objective of this study is to evaluate forage quality aspects (acid detergent fiber, neutral detergent fiber, *in vitro* digestibility, crude protein, and water soluble carbohydrates (WSC) and how they change seasonally, daily, across species and variety. This study was conducted using three sets of samples from 2014, 2015, and 2017 managed by mowing every two to four weeks to maintain vegetative growth. Samples were collected in the morning and afternoon on a monthly basis during the growing season. Flash frozen in the field, freeze dried, and ground samples were then scanned using Near Infrared Spectroscopy (NIRS) to predict forage quality and WSC content, and calibrated with lab tests. Significant species and diurnal effects were observed for WSC content. Perennial ryegrass showed the highest WSC concentrations, followed by tall fescue, Kentucky bluegrass, and orchardgrass. Afternoon harvests averaged higher WSC than morning harvests. Forage quality analysis is ongoing and will be analyzed independently and correlated to WSC. In conclusion, this research will produce important results relating the changes in forage quality, WSC, and their impact on overall equine nutrition from cool season grass pastures.

[1]Graduate Research Assistant, Plant and Soil Science, University of Kentucky (annmarie.kadnar@uky.edu); Brookdale Assistant Farm Manager, Lexington, KY; Forage Extension Specialist, Plant and Soil Science, University of Kentucky (raysmith1@uky.edu).

ASSESSMENT OF ANNUAL RYEGRASS CYTOTYPE ON CATTLE PERFORMANCE

S.M. Montgomery, J.I. Morrison and B.S. Baldwin

Annual ryegrass (*Lolium multiflorum*) can be divided into two cytotypes: diploid (2n=2x) or tetraploid (2n=4x). Past work suggests tetraploids have increased seedling vigor, more robust growth and thus a greater yield. However, there has been conflicting research suggesting no difference in yield between tetraploid and diploid varieties exists. If there is no difference in yield between cytotypes, is there a weight gain advantage in cattle grazing tetraploid cytotypes? To determine if there is an advantage in average daily gain (ADG) in steers grazing diploid and tetraploid ryegrass, a grazing study was set up at Mississippi State University Prairie Research Unit in Prairie, MS. Four common varieties, two diploid (Marshall and Tam 90) and two tetraploid (Jumbo and Nelson) were planted in 2.5-acre pastures replicated four times. Two steers were randomly assigned to each of the 16 paddocks for a 56 day grazing period. Weights were taken for each steer pre-, mid- and post-grazing throughout the season as a two-day mean. First year results indicate no difference between ADG of cattle grazing based on cytotype, but there were differences for cattle by variety. At 56-d ADG of cattle grazing Jumbo was greatest (2.45 lbs/d): followed by cattle grazing Marshall (1.92 lbs/d); Tam 90 (1.84 lbs/d) and Nelson (1.76 lbs/d). Other variables measured included: herbage mass, dry matter yield (DMY), in vitro dry matter digestibility (IVDMD), and relative forage quality (RFQ).

Masters Graduate Student, Dept. of Plant & Soil Sciences, Mississippi State University, MS (smm1006@msstate.edu); Research Associate III, Dept. of Plant & Soil Sciences, Mississippi State University, MS,(jim46@pss.msstate.edu); Professor, Dept. of Plant & Soil Sciences, Mississippi State University, MS,(bsb2@pss.msstate.edu).

EFFECTS OF FORAGE CHEMICAL COMPOSITION ON DRY MATTER DIGESTIBILITY IN THE HORSE DETERMINED USING DAISY II

V.T. Bill and L.M. Lawrence[1]

Predicting digestibility of different hay types is challenging in the horse industry due to the lack of a clear model relating forage chemical composition to digestibility. Previous forage models developed for ruminant nutrition may not apply to the horse due to differences in digestive anatomy. This study was conducted using *in vitro* methods, to evaluate the effect of chemical composition on forage digestibility. Thirty-one hay samples were used; 11 alfalfa hay samples, 13 cool-season grasses, and 7 legume-grass mixes. Forage samples ranged from 33 to 71% for neutral detergent fiber (NDF), 21 to 44% for acid detergent fiber (ADF) and 6.7 to 25.6% for crude protein (CP) (all measures presented on a dry matter basis). Samples were ground to 1-mm and weighed in triplicates into ANKOM filter bags at approximately 0.4 to 0.5 grams of sample per bag. Feces were collected from horses fed forage-based diets and were transported to the lab in a warm container to inoculate the ANKOM Daisy II vessels. Samples were incubated for either 48 or 72 h, then removed, rinsed and dried to determine dry matter digestibility. Each incubation period was repeated. Results indicate that across species, dry matter digestibility was not strongly related to CP, but decreased as NDF and ADF increased at both 48 and 72 h of incubation.

[1] Graduate Research Assistant, Department of Animal and Food Sciences, University of Kentucky, Lexington, KY (vbill14@gmail.com); Professor, Department of Animal and Food Sciences, University of Kentucky, Lexington, KY (llawrenc@email.uky.edu).

MANAGEMENTS FOR MITIGATING FESCUE TOXICOSIS

M.D. Flythe[1]

Fescue toxicosis is an ongoing problem for cattlemen and other livestock producers in regions where tall fescue is prevalent. Logically, there are two ways to alleviate or mitigate the toxicosis; either make the forage less toxic or make the animals less sensitive. Think of it in terms of four basic approaches: establish a new forage, change the physiology of the forage you have, establish new genetics in your herd or change the physiology of the animals you have.

1) Replace the tall fescue with non-toxic forage. The pastures can be reestablished with another grass or with a non-toxic endophyte tall fescue. Another lecture in this section will cover novel-endophyte fescues in detail.

2) Change the physiology of the toxic fescue. Chemical seed head suppression will maintain fescue in the vegetative state. Because the toxic alkaloid concentrations are the greatest in the seed heads, the animals will consume less toxic forage.

3) Replace sensitive animals with more tolerant animals. Since the 1990's, studies have shown that Brahman-influenced cattle are less susceptible to fescue toxicosis. Current research indicates that genotypes within British and Continental breeds can be more tolerant through differences in liver enzymes and blood vessel receptors. Another lecture in this section will discuss new data about how some cattle have rumen bacteria that metabolize ergot alkaloids differently.

4) Change the physiology of susceptible animals. Ergot alkaloid binders that purportedly prevent absorption are commercially available. However, there is still little data to show their efficacy. Isoflavones found in legumes are vaso-relaxants and counteract vaso-constriction by ergot alkaloids. Red clover has more isoflavones than other common forage legumes, but experiments also show that white clover, soybean meal and soy hulls can also improve blood flow in fescue toxicosis.

Not all strategies will be viable for all farms, but one, or a combination, is feasible in many circumstances. For example, seed head suppression in years prior to establishment of a novel endophyte fescue would mitigate toxicosis in the short term and decrease immergence of the old variety from the seed bank. Clovers could be used to improve the performance of susceptible animals while tolerant genotypes are introduced. Ongoing research by public and private scientific teams addresses each of these four strategies for mitigation or alleviation of fescue toxicosis.

[1]Research Microbiologist, USDA-ARS, Forage-Animal Production Research Unit, Lexington, Kentucky (michael.flythe@ars.usda.gov)

RUMINAL DEGRADATION OF ERGOT ALKALOIDS

B.E. Harlow and G.E. Aiken[1]

Ergot alkaloids (*e.g.*, ergovaline), produced by a fungal endophyte that infects tall fescue, can cause persistent vasoconstriction in ruminants and consequently fescue toxicosis. Ergot alkaloids are susceptible to degradation in the rumen; however, the bacteria responsible for this degradation have not been identified. This study was conducted to determine what bacteria in the rumen degrade ergot alkaloids and how adaptation to ergot alkaloid exposure impacts their degradation and consequent vasoconstriction. Laboratory experiments demonstrated that rumen hyper-ammonia-producing bacteria (HAB) degrade ergovaline and that these bacteria can adapt to increase ergovaline degradation when repeatedly exposed. Additionally, when naïve steers grazed endophyte-infected tall fescue pastures for 4 wks, some steers were able to adapt and increase rumen degradation capacity of ergovaline, but others were not able to adapt. Adaptation was specifically associated with an increase in rumen HAB and greater caudal artery luminal area, indicating that adapted steers were less constricted than steers that did not adapt. These results indicate that rumen bacteria may play an important role in ergovaline degradation and strategies for controlling their activity could have ramifications for fescue toxicosis.

Postdoctoral Fellow, USDA-ARS, Forage-Animal Production Research Unit, Lexington KY (brittany.harlow@ars.usda.gov); Supervisory Animal Scientist, USDA-ARS, Forage-Animal Production Research Unit, Lexington KY (glen.aiken@ars.usda.gov).

ADVERSE EFFECTS OF FESCUE TOXICOSIS ON ANIMAL PHYSIOLOGY

J.L. Klotz[1]

Ergot alkaloids are toxins produced by endophyte-infected tall fescue that result in a syndrome collectively referred to as fescue toxicosis. There are severe forms of this syndrome caused by high concentrations of ergot alkaloids that can manifest as 'fescue foot' where afflicted livestock suffer from near complete loss of blood flow to extremities that often times results in loss of tissue (e.g. - hooves, ear tips, tail switches). Another severe form is fat necrosis and is thought to be a result of bacterial infiltration into the body from the intestine. These two manifestations of fescue toxicosis are severe and can end with loss of the animal, but they are infrequent and often occur as management (e.g. – pasture fertilization) and environmental events converge (e.g. - weather pattern shift from drought to high precipitation conditions) into a worst-case scenario. A more prevalent and less severe form of fescue toxicosis is often referred to as 'summer slump'. Consumption of lower concentrations of ergot alkaloids over an extended period of time results in a number of effects that range from decreases in feed intake, water intake, serum prolactin, and elevated vasoconstriction. The impact of these negative effects cause decreases in average daily gain, weaning weights, and milk production, increased haircoat retention, elevated body temperatures, decreased sweating, and an increased susceptibility to heat stress. This culminates in an overall narrowing of the window in which livestock can adjust to external stressors, often times as heat stress in the warmer months. This presentation will discuss how ergot alkaloids are capable of producing these numerous and different effects and outcomes.

[1]Research Animal Scientist, USDA-ARS, Forage-Animal Production Research Unit, Lexington KY (james.klotz@ars.usda.gov).

PARADIGM SHIFT IN GRASSLAND SCIENCES: THE LIFE OF AGROSTOLOGIST AGNES CHASE

Hannah Carper, Kelsey Lainhart, and Ed Fredrickson[1]

Kuhn defined a paradigm as "an accepted model or pattern". He further describes a scientific paradigm shift as a change in the basic concepts and experimental practices of a scientific discipline. In challenging the *status quo,* Agnes Chase changed the very core of grassland sciences. As an agrostologist, Agnes Chase described the relationships among grasses and grass-like plants to provide a nomenclature and classification system that enables grassland sciences. During a time when women and men of color were prevented from achieving status as scientific professionals, Agnes Chase, and her immediate colleagues would challenge this barrier and ultimately improve how scientific data are collected, while widening the framework in which grassland sciences are conducted. We describe the life of Agnes Chase and the paradigm shift in grassland sciences she created.

[1]Department of Agriculture, Eastern Kentucky University, Richmond, Kentucky.

COOL-SEASON PASTURE RESPONSE TO MANURE TYPE AND APPLICATION DATE

Geoffrey E. Brink[1] , Michael D. Casler[1], and Wayne K. Coblentz[2]

Producers usually apply manure to pastures to improve annual productivity, but the timing of application may also influence seasonal distribution of production in the current or following growing season. Manure is typically solid, but an increasing number of pasture-based dairies utilize liquid manure. This study was conducted to determine cool-season pasture response to manure type and application date. Solid (bedded-pack; 6 T/acre) or liquid (slurry; 8000 gal/acre) dairy manure were applied to orchardgrass and meadow fescue in April, June, or August (solid only) over four years. Applied nitrogen (N) ranged from 65 to 155 lb/acre depending on the manure type and season. A non-manure paddock of each grass served as a control. Grasses were rotationally grazed throughout the growing season at vegetative-stage (12-14 in. height; 3-4 leaves) to a 4 in. residue. Across all years, orchardgrass and meadow fescue had equivalent annual production (5540 lb DM/acre) and efficiency of N use (13.4 lb DM/lb applied N). After the second rotation, meadow fescue always had lower fiber and greater fiber digestibility than orchardgrass. Seasonal distribution of pasture production was not influenced by manure type or application date; relative yield of both grasses declined after the spring, although the decline was greatest in meadow fescue. Grasses fertilized with liquid manure produced 3 lb more DM/lb applied N than solid manure, and spring application of either manure produced 5 lb more DM/lb applied N than summer or fall application. Although the option of applying manure in summer or fall offers some management flexibility, spring application provides the best production return on the N applied.

[1] Research Agronomist and Research Geneticist, respectively, USDA-ARS, U.S. Dairy Forage Research Center, Madison, WI 53706; [2] Research Dairy Scientist, USDA-ARS, Institute for Environmentally Integrated Dairy Management, Marshfield, WI 54449.

DEVELOPMENT OF A FARMER-BASED WORKING GROUP PROGRAM FOR BEEF-FORAGE SYSTEMS MANAGEMENT IN ALABAMA

G.L. Thompson, and M.K. Mullenix[1]

A working group is defined as a farmer-to-farmer network focused on a common theme or series of topics of interest. In 2016, a farmer-based working group was organized in northern Alabama to discuss beef-livestock management within the region. A series of five meetings were organized to highlight technologies and strategies for agronomic, animal, and economic best practices in beef cattle operations. Farmer meetings were located on-farm or at Auburn University affiliated research stations and demonstration sites, and offered from August 2016 through April 2017 to illustrate different management seasons. Topics included value-added calf marketing, bull selection, precision soil sampling, rotational grazing, bale grazing, and water quantity and quality in grazing systems. Group discussion were led by farmer participants, personnel from Alabama Extension, USDA NRCS Alabama, or another regional land-grant institution. A survey was conducted at the final meeting in April 2017 (n = 26 participants; average 66 head of cattle and 109 acres per operation). 100% of the participants indicated that the program met their expectations, and that it should be offered again. 75% of the farmers in the group reported that they had already adopted at least one of the management topics discussed as part of the program. Specifically, forage management strategies adopted by these farmers included working towards more than 300 days of grazing per year, precision soil sampling, and improving grazing management strategies. Participants reported an average economic impact of $2,500 per operation.

[1] Animal Science and Forage Regional Extension Agent, Alabama Cooperative Extension System, Auburn, AL (thompgl@auburn.edu); Assistant Professor/Extension Specialist, Alabama Cooperative Extension System/Auburn University, Auburn, AL (mullemk@auburn.edu);

GROWTH RATES OF COOL-SEASON GRASS-LEGUME CANOPIES DEFOLIATED AT A RANGE OF LIGHT INTERCEPTION LEVELS

T.C. Griggs, M.L. Satterfield, J.M.A. Koos, Z.B. Freedman, and C.N. Kelly[1]

Pasture growth rate varies with species physiology and structure, developmental stage, environmental conditions, and timing and severity of defoliation. Relationships of growth rate and total-season herbage production to canopy light interception (LI) and corresponding leaf area index (LAI) across defoliation stages including late maturity are not well-defined for cool-season mixed-species swards in the Appalachian region. We hypothesized that pasture growth rate and total-season forage production will be less for relatively few harvests of late-maturity hay-stage pasture at >95% LI (LAI>6-8) than for more-frequent harvests at 90-95% LI (LAI <4-6). We compared LI and growth rate of perennial pastures containing cool-season grasses, legumes, and non-legume forbs under defoliation frequencies and residual heights simulating continuous and rotational stocking conditions ranging from a) weekly to 3 or 5 in; b) periodically from 10-12 in to 2.5 or 5 in; and c) periodically from mature hay stage to 2.5 or 5 in. Herbage mass was estimated with a calibrated rising plate meter and LI and LAI were estimated from canopy light interception measurements with a line quantum sensor. Herbage mass and LI levels during regrowth periods and responses of growth rate and total-season herbage production to LI varied among defoliation treatments will be presented.

[1]Forage Agronomist, Div. of Plant and Soil Sciences (tcgriggs@mail.wvu.edu); Research Assistant, Div. of Plant and Soil Sciences (msatterf@wvu.edu); Graduate Research Assistant, Div. of Plant and Soil Sciences (jmk0035@mix.wvu.edu); Soil Microbiologist, Div. of Plant and Soil Sciences (zachary.freedman@mail.wvu.edu); Soil Scientist, Div. of Forestry and Natural Resources (ckelly1@mix.wvu.edu), West Virginia University, Morgantown, WV.

BALANCING MIXED RATIONS WITH GRAZING USING NIRS FOR PASTURE ANALYSIS

T.C. Griggs[1]

Pastures can present limitations to daily dry matter (DM) and nutrient intake by grazing animals. Energy supplementation is commonly used to maximize pasture utilization while meeting more of the nutritional demands of high-producing dairy cattle than can be provided by pasture alone. Partial mixed rations that complement pasture nutrient intake to support a target level of milk production can be challenging to formulate because of temporal and spatial variability of pasture mass and composition, excessive levels of protein and protein degradability in pasture herbage, and inaccuracies in predicting DM and nutrient intake by grazing animals. Pasture DM intake remains one of the most difficult nutritional parameters to predict accurately from herbage compositional information. While near-infrared reflectance spectroscopy (NIRS) is widely used for determination of the composition of dried, ground forage samples, proximal analysis of intact pasture canopies is a relatively undeveloped approach that can improve predictions of daily DM and nutrient intake. Proximal analysis refers to remote sensing with hand-held or other ground-based spectral sensors. Although numerous problems remain to be resolved, there is increasing evidence for the possible role of proximal analysis of pasture canopies. This presentation will review pasture compositional parameters, including DM intake, that may be readily evaluated via field spectroscopy to support formulation of mixed rations for pasture-based dairy production.

[1]Forage Agronomist, Div. of Plant and Soil Sciences, West Virginia University, Morgantown, WV (tcgriggs@mail.wvu.edu).

GENETIC ANALYSIS OF *CYNODON DACTYLON* FORAGE CULTIVARS USING SEQUENCE-RELATED AMPLIFIED POLYMORPHISM MARKERS

J.C. Hitzelberger[1], R.A. Lane[1], C.P. Randle[2]

Interpretive Summary

Cynodon dactylon is a warm season perennial grass native to the African continent. It was likely introduced to the U.S. in the 1700s, and was documented as an important forage grass in the U.S. by 1807 (USDA Fact Sheet, 2000). Now grown across most of the southern United States, it is widely utilized for turf and forage purposes, occupying more than 10 million acres (Burton and Hanna, 1995). Forage types of bermudagrass are grown for use in hay production and grazing livestock. Bermudagrass cultivars vary in biomass production, stem and leaf size, disease resistance, and stress tolerance. The most popular forage cultivars in use today are hybrids and are sterile. This means they must be propagated asexually (cloned) from vegetative tissues, such as rhizomes, stolons, and vertical stems.

C. dactylon is desired for its tolerance to environmental stresses, biomass production, and nutritional quality for livestock production. While able to survive on low nitrogen soils, it requires high nitrogen fertilization to achieve optimum dry matter yield and nutritional value, a major criticism against its sustainability. Tolerance to heavy grazing, heat, salinity, and drought make it a strong choice for Texas hay and cattle producers. Resistance to common fungal diseases such as rust (*Puccinia graminis)* and *Helminthosporium* leafspot varies in cultivars. These diseases cause major yield and stand loss. Disease resistance is a valued trait in *C. dactylon.* Other trait improvements are continually sought to improve sustainability, nutritional value, palatability and stress tolerances.

Cynodon dactylon cv. Raven (Raven) is a recent accession from the Sam Houston State University campus in close proximity to Raven Nest Golf Course on Interstate 45. The area was the previous site of the Sam Houston State University dairy farm and has no fertility or irrigation management in the non-playable "rough" area of the course. The dairy pasture was possibly planted with *C. dactylon* cv. *Coastal* (Coastal), a sterile hybrid, and *C. dactylon* var. *dactylon* (Common) a seed producing variety, though there is no record of either being purposely planted there. Raven shows phenotypic variation from Common and Coastal. Raven produces no viable seed, suggesting it is a result of natural hybridization. The stem and leaf size of Raven is considerably larger than Common and slightly larger than Coastal (Lane, personal observation). Upon accession, Raven was visibly dense and healthy even with poor fertility and lack of irrigation, suggesting a greater level of sustainability. Under identical production practices, Raven has comparable biomass production and nutritional quality to current *C. dactylon* forage type cultivars and is less susceptible to foliar fungal diseases than Jiggs (Unpublished data,

[1]Department of Agricultural Sciences, Sam Houston State University, Huntsville, Texas 77341-2088.
[2]Department of Biological Sciences, Sam Houston State University, Huntsville, Texas 77341-2088.

Lane). Vegetative tissue of Raven was genetically analyzed and compared to four other commercially available forage cultivars as well as other accessions from Raven Nest Golf Course. The four cultivars compared in this study include: *Cynodon dactylon* var. *dactylon, C. dactylon* cv. *Coastal, C. dactylon* cv. *Tifton 85, C. dactylon* cv. *Jiggs.*

A total of ten (10) tissue samples were collected for genetic analyses. Two accessions of *C. dactylon* cv. *Jiggs* (Jiggs) were compared. The two accessions were from the same hay field but were planted at separate times (2000 and 2010) and from different source material. Jiggs bermudagrass is a farmer-selected cultivar. The plant material sources of both Jiggs plantings are established at the SHSU Gibbs Ranch.

Three unknown *C. dactylon* accessions from Raven Nest Golf Course were also collected that vary slightly from Raven phenotypically. These accessions are referred to in relation to where they were found on the course; Hole 1, Hole 16 and Hole 18, all in the "rough" area. One accession of Raven was collected from the original accession location on Raven Nest Golf Course and another was made from a trial plot established off-location in 2010 to compare relatedness of the collections. Tifton-85, Common, and Coastal were used to ensure a complete analysis of possible relatedness of Raven to existing cultivars.

Sequence-Related Amplified Polymorphism primers produced polymorphic markers. The SRAP marker analysis proved useful for the genetic testing and comparison between *Cynodon dactylon* forage cultivars used in the southern United States. After generating a Principle Coordinate Analysis map, it can be concluded *Cynodon dactylon cv. Raven* is genetically different from the other forage cultivars tested. Raven showed the greatest distance from Tifton 85, Coastal, and Common. Both Raven accessions clustered with Hole 1, Hole 16, and Hole 18 accession.

Jiggs was the most closely related cultivar to Raven. The two samples of Jiggs were slightly different; as were the two samples of Raven. This could be due to small variations within the genomes or error within the protocol or background noise associated with mismatched primer alignment. Raven is far enough removed genetically from Jiggs and other tested *C. dactylon* forage grasses to suggest genetic uniqueness. The sterile nature of Raven suggests a wild hybridization is most likely to have occurred in that area. The time at which the hybridization took place is unknown but could explain the relatedness and the diversity within the cluster. Further research with more precise markers could be done to determine hybridization events and pedigrees.

ASSESSING THE EFFECT OF GIBBERLLINS ON RYE AND ANNUAL RYEGRASS FORAGE SYSTEMS IN THE SOUTHEASTERN USA.

Jessica Long[1]and Dennis Hancock· PhD[2]

Plant hormones stimulate or inhibit certain responses by the plant, inducing physiological changes. Plant growth regulators are chemicals designed to enhance the effect of a certain class of hormones. Gibberellins are a class of plant hormone that stimulate stem and leaf elongation and plant growth regulators that contain gibberellins may induce higher yield. The objective of this study was to assess the use of the gibberllin, GA_3, applied to cereal rye (*Secale cereale*) and annual ryegrass (*Lolium multiflorum)* in the fall (F), winter (W), and fall + winter (FW) application over the course of three growing seasons (2011-12, 2012-13, and 2013-14) to determine the effect on yield and forage quality. The results showed that GA_3 increased forage yield for each application date in the first year of the study and in response to the FW application in all three years. In 2013 and 2014, the nutritive value (e.g., TDN, CP, RFQ, etc.) of treated forage was not affected by the timing of treatment. In contrast, the forage that received a W application in the 2011-12 growing season resulted in lower average TDN, CP, and RFQ and higher ADF, NDF, and lignin than that receiving a F or FW application. This study shows that GA_3 has the potential to increase forage yield, especially when receiving a fall and winter application. Further, the application of GA_3 has minimal impact on the nutritive value of rye and/or annual ryegrass.

[1]Graduate Student-University of Georgia, Athens GA; [2]Forage Extension Specialist-University of Georgia, Athens GA.

EFFECT OF N, P, AND K FERTILIZATION ON YIELD OF NATIVE WARM-SEASON GRASSES

J.D. Richwine[1], P.D. Keyser[2], A.J. Ashworth[3], and G.E. Bates[4]

Fertility responses of native warm-season grasses in a forage production setting have not been well documented. Three studies were established at East Tennessee AgResearch and Education Center in Knoxville, TN, in 2013 to evaluate yield response of switchgrass (SG) and big bluestem (BB) to nitrogen (N; 0, 30, 60, 120, 180, and 240 lb/ac), phosphorous (P; 0, 60, 90, and 180 lb/ac), and, for SG only, potassium (K; 0, 75, 150, and 275 lb/ac). All fertilizers were applied in two split-applications with the first at green-up (late April) and the second following the first harvest (mid to late June). Plots were harvested to an 8-inch stubble height twice each calendar year, 2014-2016. For N, SG at 240 lb/ac (4.56 T/ac) and BB at 180, 120, and 240 lb/ac (4.43, 4.40, and 4.07 T/ac, respectively) produced comparable yields across all years. For SG, the only N rate not significantly greater than 0 lb/ac (2.27 T/ac) was 30 lb/ac (2.50 T/ac.). For BB, all rates (3.03-4.43 T/ac) of N were greater than 0 lb/ac (2.04 T/ac). For P and K, yield was not influenced. These results indicate N is the most influential nutrient for SG and BB with respect to yield for any given year.

[1]Graduate Research Assistant, Center for Native Grasslands Management, Dept. of Forestry, Wildlife, and Fisheries, University of Tennessee (UT), Knoxville, TN (jondrich@vols.utk.edu); [2]Professor, Center for Native Grasslands Management, Dept. of Forestry, Wildlife, and Fisheries, UT, Knoxville, TN (pkeyser@utk.edu); [3] Research Scientist, USDA-ARS, Fayetteville, AR (Amanda.Ashworth@ARS.USDA.GOV); [4]Professor, Dept. of Plant Sciences, UT, Knoxville, TN (gbates@utk.edu).

WHICH NATIVE WARM-SEASON GRASS VARIETY WOULD YOU PLANT?

J.D. Richwine[1] and P.D. Keyser[2]

Big bluestem [*Andropogon gerardii* Vitman] and Indiangrass [*Sorghastrum nutans* (L.) Nash.] are forages well adapted to the Mid-South, but biomass yield of commercially available varieties of these species for forage production has not been well documented. Therefore, we established a variety trial at East Tennessee AgResearch and Education Center in Knoxville, TN, in 2013 to evaluate seven big bluestem and six indiangrass varieties, 2014-2016. Two big bluestem varieties, 'Karst' (5.42 T/ac) and 'Mammoth' (5.39 T/ac), produced greater ($p = 0.0002$) biomass than four out of five remaining big bluestem varieties (four varieties ranging from 3.29-4.22 T/ac). There was no difference among the six indiangrass varieties ($p = 0.3097$; 3.22-4.22 T/ac). Big bluestem varieties showed no pattern from year to year while IG varieties remained consistent with respect to yield. Forage quality analyses are pending for these 13 entries.

[1]Graduate Research Assistant, Center for Native Grasslands Management, Dept. of Forestry, Wildlife, and Fisheries, University of Tennessee – Knoxville, Knoxville, TN (jondrich@vols.utk.edu); [2]Professor, Center for Native Grasslands Management, Dept. of Forestry, Wildlife, and Fisheries, University of Tennessee – Knoxville, Knoxville, TN (pkeyser@utk.edu).

ALABAMA ECOLOGY OF GRAZING LANDS SHORT COURSE: AN INTERAGENCG TRAINING EFFORT ON IMPROVED FORAGE MANAGEMENT PRACTICES

J.B. Elmore[1], M.K. Mullenix[2], M.K. Stanford, J.B. Gladney[3], G.R. Thompson[4], and S. Weaver[5]

A series of workshops related to forage ecosystem management and sustainability were held from May 2016 through May 2017 for USDA Natural Resources Conservation Service – Alabama (NRCS) personnel and Alabama Cooperative Extension System (ACES) extension agents in Alabama. The goal of these workshops was to provide practical learning opportunities related to forage management for technical service providers with these agencies. In-classroom and hands-on demonstrations illustrated grazing management concepts, soil-plant animal evaluation practices, and facilitated group problem-solving and discussion. Attendees (n=67) reported that outdoor, hands-on demonstrations related to grazing management concepts were the most effective in improving their knowledge of forage practices (68%). Field demonstrations on pasture assessment, grazing methods, and silvopasture were listed as the most useful to participants (56%). Participants were highly likely (65%) or likely (35%) to share these practices with farmers during site visits and one-on-one consultation with clientele in the next 12 months following these programs. Results indicate that joint in-service training opportunities among university and government organizations may benefit communication, understanding, and application of forage management practices in on-farm settings.

[1]Regional Extension Agent Alabama Cooperative Extension System, Clanton, AL (elmorjb@aces.edu); [2]Assistant Professor/Extension Specialist, Animal Sciences, Auburn University/Alabama Cooperative Extension System, Auburn, AL (mullemk@auburn.edu); [3]Extension Specialist, Crop, Soil and Environmental Sciences, Auburn University/Alabama Cooperative Extension System, Crossville, AL (stanfmk@aces.edu); [4]Regional Extension Agent, Alabama Cooperative Extension System, Tuscaloosa, AL (gladnjb@aces.edu); [5]Regional Extension Agent, Alabama Cooperative Extension System, Auburn, AL (thompgl@aces.edu); [5]Assistant State Conservationist for Technology, Natural Resource Conservation Service, Auburn AL (Shannon.Weaver@al.usda.gov).

IT WORKS! ANNUALS FOR HAYLAGE

J. Yates, C. Vaught, K. J. Simon, and J. A. Jennings[1]

In Southwest Arkansas, many livestock producers get in the habit of "chasing hay" all over the state. Our goal was to produce quality hay in a small amount of time and acreage. The winter annual forage species included Elbon rye, winter wheat, Jerry oat, Gulf ryegrass. Plots were prepared, tilled, and planted on a double-cropped site. Plots were planted March 23, 2017. The 6.6-acre was used as a corn maze in the previous fall. Prior to planting, weeds were sprayed with the non-selective herbicide glyphosate. All the plots were harvested on May 17, 2017. Yield for Gulf ryegrass was 2.86 tons/acre, Elbon rye was 2.0 tons/acre, winter wheat was 1.53 tons/acre, and Jerry oat was 4.2 tons/acre. Jerry oat was the top performer. The forage was wrapped into haylage.

[1]County Extension Agent - Agriculture, University of Arkansas, Cooperative Extension Service, Mean, AR (jjyates@uaex.edu); [1]County Extension Agent - Agriculture, University of Arkansas, Cooperative Extension Service, Mean, AR (cvaught@uaex.edu) Animal Science Program Associate – Forages (ksimon@uaex.edu), Professor and Extension Forage Specialist (jjennings@uaex.edu) , and Professor (pbeck@uaex.edu) University of Arkansas Division of Agriculture. Little Rock, Arkansas. 72204.

EVALUATION OF HERBICIDE-IMPREGNATED DRY FERTILIZER FOR WINTER ANNUAL BROADLEAF WEED CONTROL

K. M. Davis[1], B. Griffin[1], K.J. Simon[2], J.A. Jennings[2] and B.Williams[3]

Winter annual broadleaf weeds can overtake a pasture when left untreated. Proper weed control requires an integrated approach of herbicide and fertility management. Due to time constraints or lack of equipment, many forage producers fail to do both. In 2017, GrazonNext HL received a label in AR for mixing with dry fertilizer allowing a producer to control broadleaf weeds and promote forage growth in one pass across the pasture. A replicated on-farm research trial was conducted for broadleaf weed control featuring seven treatments to evaluate the effectiveness of conventional foliar sprayed herbicides and separately applied dry fertilizer versus herbicide-impregnated dry fertilizer. Fertilizer was applied according to soil test recommendations for warm-season grass pasture. P and K were applied with the herbicide treatments on March 20. N was applied a month later. Herbicides were applied at 1 qt/ac with 0.25% nonionic surfactant using a boom sprayer. Treatments were evaluated on day 21 and again on day 70. We found that GrazonNext HL impregnated dry fertilizer had 80% control of winter annual broadleaf weeds, while foliar applications of all herbicides ranged from 90-98% control. 2017 was an optimal year for applying GrazonNext HL impregnated dry fertilizer, as rainfall is needed to activate the herbicide. This is a great option for small weeds, less than 6 inches in height.

[1]County Extension Agent - Agriculture, University of Arkansas, Cooperative Extension Service, Nashville, AR (kmdavis@uaex.edu); CEA - Agriculture, UACES, Clarksville, AR (bgriffin@uaex.edu); [2]Animal Science Program Associate - Forages, UACES, Little Rock, AR (ksimon@uaex.edu); Animal Science Professor and Extension Forage Specialist, UACES, Little Rock, AR (jjennings@uaex.edu); [3]Range and Pasture Specialist, Dow AgroSciences, Memphis, TN (bwwilliams@dow.com).

IMPACTS OF WINTER SUPPLEMENTATION OF GROUND WHOLE FLAXSEED ON MILK FATTY ACID COMPOSITION ON ORGANIC DAIRY FARMS IN THE NORTHEASTERN UNITED STATES

Aimee N. Hafla, Kathy J. Soder, André F. Brito, Richard Kersbergen, A. Fay Benson, Heather M. Darby, Melissa D. Rubano, S. Leanne Dillard, Jana Kraft, and Simone F. Reis[1]

The objective of this study was to evaluate supplementation of ground whole flaxseed to organic dairy cows during the non-grazing season to maintain levels of beneficial fatty acid concentrations in milk typically observed during the grazing season. During the winters of 2013-14 and 2014-15 (non-grazing season), 9 organic dairy farms supplemented ground whole flaxseed at 6% of diet dry matter to half of the cows within each herd. Milk samples were collected monthly throughout the 90-day feeding period. Winter flaxseed supplementation did not impact concentrations of milk fat or milk protein. While flaxseed increased milk omega-3 fatty acid concentrations by 88%, total conjugated linoleic acid (CLA) concentrations in milk were only slightly increased (9%). A greater level of winter supplementation is required to increase concentrations of beneficial fatty acids, such as CLA, compared to concentrations observed during the grazing season. However, due to potential negative impacts on ruminal fermentation and milk production at greater levels of flaxseed supplementation, economic incentives would be needed to justify feeding flaxseed to dairy cows to improve milk fatty acid profiles.

[1]Animal Scientists, USDA-ARS, PSWMRU, University Park, PA (aimee.hafla@gmail.com; Kathy.Soder@ars.usda.gov; melissa@rreng.net; dillardsa@auburn.edu); Animal and Nutritional Sciences, University of New Hampshire (Andre.Brito@unh.edu, frotassimone@gmail.com); University of Maine Cooperative Extension (Richard.Kersbergen@maine.edu), Cornell Cooperative Extension (afb3@cornell.edu), Plant and Soil Science, University of Vermont (Heather.Darby@uvm.edu), Animal and Veterinary Sciences, University of Vermont, (Jana.Kraft@uvm.edu).

OVERSEEDING EASTERN GAMAGRASS: AN ALTERNATIVE FORAGE SYSTEM FOR THE BLACK BELT REGION

K.M. Mason[1], M.K. Mullenix[1], J.J. Tucker[2], J.S. Angle[3], R.B. Muntifering[1], and J. Yeager[4]

Overseeding Eastern gamagrass (EG) with cool-season forages may provide an alternative to tall-fescue based forage production systems in the Alabama Black Belt; however, little is known about potential effects on native warm-season grass (NWSG) system sustainability. A two-year study was conducted to determine the effects of overseeding with cool-season forage mixtures and grazing management on forage production, nutritive value, and persistence of Eastern gamagrass. During the cool season months, NWSG were overseeded in small plot study as follows: Control (no overseeding), Florida 401 rye, Wrens Abruzzi rye, and Wrens Abruzzi Rye + AU Red Ace Red Clover. During the warm season, NWSG received one of two N fertility treatments (60 lb N/ac or 120 lb N/ac). Mob stocking was used to simulate rotational grazing every 28 days and forage samples were clipped at each grazing event to determine herbage mass and nutritive value. Cover ratings were taken during transition months at the beginning and end of the NWSG production season to determine EG persistence. Results indicate that small-grain mixtures may fit well in a overseeding system with NWSG based on growth distribution, and provide winter grazing. Overseeded grasses and clover can provide ground cover to reduce weed pressure in the spring to summer transition. Overseeding did not affect EG herbage mass and persistence, though forage production was relatively low.

[1]Department of Animal Sciences, Auburn University, Auburn, AL (kmb0108@auburn.edu), (mullemk@auburn.edu), (muntirb@auburn.edu); [2]Department of Animal and Dairy Science, University of Georgia, Tifton, GA (jjtucker@uga.edu); [3]International Fertilizer Development Center, Muscle Shoals, AL (jangle@ifdc.org); [4]Black Belt Research and Education Center, Auburn University, Marion Junction, AL (yeagejj@auburn.edu)

GRAZING STOCKERS ON DIVERSE MIXTURES OF SUMMER ANNUAL FORAGES

K.M. Mercier, C.D. Teutsch, S.R. Smith, E.L. Ritchey, K.H. Burdine, and E.S. Vanzant[1]

Many fall-calving cow-calf producers sell weaned calves in the springtime. This study was conducted to evaluate the economic and ecologic impacts of grazing weaned calves on summer annual pastures instead of selling calves at weaning time. Three summer annual treatments included 1) a monoculture of sorghum-sudangrass, 2) a simple mixture of sorghum-sudangrass, pearl millet, and soybeans, and 3) a complex mixture of sudangrass, corn, crabgrass, sunn hemp, sunflower, cowpea, Korean lespedeza, and two types of brassicas added to the simple mixture components. Seven stocker calves (726 lbs. initial weight) and 3-4 cows strip-grazed approximately 4-acre paddocks from June 13 until forages ran out in August 2017. Cumulative forage production per acre (measured prior to last grazing event) was not different between summer annual treatments ($p < 0.05$). Calves grazing simple mixtures gained an average of 1.8 lbs/day, followed by calves grazing monocultures (1.7 lbs/day), and complex mixtures (1.5 lbs/day). Forage maturity most likely played a large role in low average daily gains which will impact the economic feasibility of these systems. Forage quality, botanical composition, and forage utilization data will also be presented.

[1]Graduate Student, University of Kentucky Research and Education Center at Princeton, Princeton, KY (kelly.mercier@uky.edu); Extension Associate Professor, University of Kentucky Research and Education Center at Princeton, Princeton, KY (chris.teutsch@uky.edu); Extension Professor, University of Kentucky, Lexington, KY (raysmith1@uky.edu); Extension Associate Professor, University of Kentucky Research and Education Center at Princeton, Princeton, KY (elritc2@uky.edu); Associate Extension Professor, University of Kentucky, Lexington, KY (kburdine@uky.edu); Associate Professor, University of Kentucky, Lexington, KY (evanzant@uky.edu).

FINISHING FORAGES FOR GRASS-FED BEEF

K.A. Cassida, R.M. Martin, J.E. Rowntree, J. Schweihofer, and M. Raven[1]

One of the most challenging periods in grass-fed beef production is the last few weeks before harvest, when cattle need the best possible pasture to put on the last few pounds of finish. When using a spring calving season and aiming for harvest at 19 months of age, this time coincides with end of growing season decline in forage availability. Cold-tolerant, highly digestible forage brassicas may offer a solution to this dilemma. In central Michigan, we compared forage yield and quality in replicated (n=4) mixed perennial grass-legume pasture with simple (forage rape and oats) and complex (two varieties of rape, oats, field peas, and Italian ryegrass) annual mixtures. Annuals were planted early-June 2014-16 and strip-grazed for approximately ten weeks beginning in late August. During the same period, stockpiled perennial pasture was grazed using a rotational stocking system. Each pasture contained two finishing steers with initial live weight averaging 960 lb. Forage mass was hand-clipped, quantified, and sampled for nutritive value analysis once per week before and after grazing. Averaged over all years, pre-grazing forage biomass and crude protein content was greater for the two annual forage treatments than for perennial grass/legume (4550 vs. 2790 lb/acre, $P < 0.05$). Crude protein was relatively low but greater in complex annual and perennial pastures than in simple annuals (12.3 vs 11.1%, $P < 0.05$). There were complex treatment by week interactions for cell wall components, digestibility and ash concentration of forage.

[1]Fixed Term Assistant Professor, (cassida@msu.edu); Graduate Student (mart1695@msu.edu); Associate Professor (rowntre1@anr.msu.edu), Extension Educator (grobbelj@msu.edu), and Professor (mraven@anr.msu.edu), Michigan State University, East Lansing, MI.

YIELD AND QUALITY OF REDUCED LIGNIN ALFALFA ACROSS SIX STATES

K. Cassida, R.M. Sulc, A. Parker, K. Albrecht, M. Hall, D. Min, S. Orloff, and D.Undersander[1]

Harvest timing of alfalfa is critical because lignification of cell walls with advancing maturity rapidly reduces nutritive value. Alfalfa with the recently released transgenic reduced-lignin trait should maintain acceptable nutritive value longer than conventional alfalfa cultivars, thus allowing more flexibility in harvest timing. We compared forage yield and nutritive value of HarvXtra™ alfalfa to two conventional cultivars (high yield potential cv. 54R02 and high nutritive value cv. WL355RR) at three harvest intervals (28, 33, or 38 days) in six environments across the USA (CA, KS, MI, OH, PA, WI) during the seeding year and year after seeding. When HarvXtra™ was compared with the mean of 54R02 and WL355RR across two cuttings for all sites in both years, it consistently contained less acid detergent lignin and had greater neutral detergent fiber digestibility (NDFD) than the other cultivars. As expected, NDFD decreased as harvest interval increased. Across sites and cuttings, HarvXtra™ yield was slightly lower than for the other cultivars. Transgenic reduced-lignin alfalfa could be harvested 5 to 10 days later with similar nutritive value as the other cultivars harvested 5 to 10 days earlier in most harvest environments. In certain regions, this could translate into HarvXtra™ being harvested one time less per year than the other cultivars, translating into reduced costs and potentially improving stand persistence of the alfalfa.

[1]Cassida, Fixed Term Assistant Professor, (cassida@msu.edu), Michigan State University, East Lansing, MI; Albrecht, Professor (kaabrec@wisc.edu) & Undersander, Professor, (dunders@wisc.edu), University of Wisconsin-Madison, Madison, WI; Hall, Professor, (mhh2@psu.edu), Pennsylvania State University, State College, PA; Min, Assistant Professor, (dmin@ksu.edu), Kansas State University, Manhattan, KS; Orloff, UCCE Siskiyou County, Yreka, CA; Sulc, Professor (sulc.2@osu.edu) and Parker, graduate student (parker.773@buckyemail.osu.edu), The Ohio State University, Columbus, OH.

EVALUATION OF FERMENTATION ENHANCERS ON FORAGE PRESERVATION CHARACTERISTICS IN ANNUAL RYEGRASS BALEAGE IN ALABAMA

M.K. Mullenix, L.D. Roth, J.B. Elmore, M.E. Griffin[1]

Use of annual ryegrass for baleage production in the Southeast US represents a potential high-quality forage resource for beef cattle producers. Few studies have evaluated the effect of fermentation inoculants on storage preservation of cool-season baleage. The objective of this study was to evaluate two fermentation enhancers (Forage-Mate VS-3™ or Promote® HayDefender™, Cargill Animal Nutrition) and time of wrapping after baling on forage dry matter (DM), nutritive value, and bale temperature following a 90-d storage period. Final bale weight was less for delay-wrapped bales than those wrapped the same day as baling (mean 1025 vs. 1140 lb DM/bale, respectively). No differences were observed among forage inoculants in final bale weight. Forage CP, NDF, ADF, and ADICP did not differ among forage inoculants or time of wrapping treatments. Overall, IVTD of annual ryegrass baleage was greater than 80% in this study across inoculant types. Delayed wrap time increased average bale temperature more rapidly when bales were opened at 90-d compared to those wrapped the same day. Control and VS-3-treated bales had greater temperature than HayDefender bales after opening over a simulated three-day feeding period. HayDefender treated bales were more stable in terms of temperature at the time of feeding. Results indicate that time of wrapping had a greater impact on bale weight and temperature regulation than forage inoculants in annual ryegrass baleage.

[1]Assistant Professor/Extension Specialist, Animal Sciences, Auburn University/Alabama Cooperative Extension System, Auburn, AL (mullemk@auburn.edu); Research Nutritionist, Provimi North America, Cargill Animal Nutrition, Minneapolis, MN; Animal Science and Forage Regional Extension Agent, Alabama Cooperative Extension System, Auburn, AL (elmorjb@auburn.edu); Graduate Research Assistant, Animal Sciences, Auburn University, Auburn, AL (mzg0054@auburn.edu).

PASTURE MANAGEMENT TO OPTIMIZE HORSE HEALTH

Bridgett McIntosh[1]

Pasture provides horses with an excellent source of nutrition, as well as opportunities for exercise and interaction with other horses, optimizing overall health. While properly managed pastures can provide the cornerstone of the equine diet, fluctuations in forage nonstructural carbohydrates (NSC) can be detrimental to horse health. Elevated intake of forage NSC by grazing horses can lead to laminitis as a result of altered metabolic, digestive, and/or inflammatory response. Laminitis is a painful, disabling and often deadly inflammatory disease culminating in the laminae of the hoof. Obese horses and those with known metabolic dysregulation are at greatest risk for developing pasture associated laminitis. Improved understanding of the relationships between the environment, forages, the horse and subsequent management techniques are necessary to mitigate risks for laminitis. In studies conducted in northern Virginia, circadian and seasonal variation occurred in forage NSC in a mixed cool season grass pasture under continuous grazing. Environmental factors including sunlight, temperature and humidity were associated with changes in forage NSC content. Weekly and seasonal fluctuations in metabolic, digestive and inflammatory responses of horses were in turn related to forage NSC content. A rotationally grazed pasture at the same site had overall higher concentrations of NSC compared to a continuously grazed pasture. Grazing strategies that reduce the intake of pasture NSC are useful in mitigating the risk of pasture associated laminitis in horses.

[1]Equine Extension Specialist, Middleburg Agricultural Research and Extension Center, Department of Animal and Poultry Sciences, Virginia Tech, Middleburg, Virginia. (bmcintosh@vt.edu)

GRAZING BERMUDAGRASS IN KENTUCKY: LESSONS LEARNED

Bob Coleman[1]

A grazing trial was conducted over 3 years to evaluate the use of bermudagrass in horse pastures. The aims of the project were to determine bermudagrass grazing strategies and if the forage could maintain mature horses. A 4.8 acre pasture was established by seeding 'Wrangler' bermudagrass two years prior to grazing. The pasture was fertilized with 250 pound of nitrogen per acre at the start of each grazing season. Mature thoroughbred mares at maintenance with an average body condition score (BCS) of 5.5 were used for grazing (Table 1). Mares were rotationally grazed when bermudagrass was grazed down to an average height of 3 inches, resulting in grazing periods of 5 to 9 days and rest periods of 21 to 30 days. Post-grazing, pastures were mowed to 3 inches and dragged to scatter manure. Horses at maintenance were able to meet their nutritional requirements while grazing bermudagrass with minimal changes in bodyweight (BW). The stocking rate and single species allowed for uniform grazing across the pasture, reducing the formation of lawns and roughs. It was estimated that horses consumed 45 to 50% of the available bermudagrass. Bermudagrass pasture provided ample forage and horses maintained BW and BCS with no supplemental feed. Management of bermudagrass pastures should include both fertilization and mowing if incorporated it into a horse grazing system.

Table 1. Characteristics of Thoroughbred mares grazing a bermudagrass pasture in Kentucky.

Yr.	No. horses	Initial BW (lb.)	Ending BW (lb.)	BW change (%)	Stocking rate*	Grazing days
1	11	1,351	1,347	-0.28	2.3	109
2	14	1,296	1,265	-2.5	2.9	59
3	14	1,274	1,270	-0.13	2.9	82

*Horses per acre

[1]Associate Professor, Department of Animal and Food Sciences, University of Kentucky, Lexington, Kentucky. (rcoleman@email.uky.edu)

MANAGING HORSE PREFERENCE IN PASTURES

Laurie Lawrence[1]

Horses are known as selective grazers. Pastures grazed by horses frequently develop a "roughs and lawns" appearance, with the lawns representing the more heavily grazed areas and roughs representing the less heavily grazed areas. In many cases, the lawns can be rapidly overgrazed, reducing the longevity and productivity of the pasture plants in those areas. Improved understanding of the factors affecting horse grazing preferences may enhance pasture management. In cool-season grass pastures, horses demonstrate preferences for different grasses and even within a species they may demonstrate preferences for certain cultivars. In Central Kentucky studies conducted at the beginning of the spring grazing season, horses discriminated against tall fescue cultivars compared to other cool season grasses. We did not observe that preferences were related to concentrations of protein or soluble carbohydrates. We also did not observe any difference in the preference of horses for endophyte-free versus endophyte-infected tall fescue. Forage persistence was inversely related to preference ranking for most cultivars. We hypothesize that horses graze preferred forages more intensely, which compromises the persistence of the forage within the pasture. It is possible that seeding pastures with grasses that have high preference scores may lead to shorter plant longevity and that use of cultivars with low preference scores may result in pastures with longer persistence.

[1]Provost's Service Professor, Equine Nutrition, Department of Animal and Food Sciences, University of Kentucky, Lexington, Kentucky (llawrenc@uky.edu).

GRAZING ALTERNATIVE FORAGES IN HORSE PASTURES

Krishona L. Martinson[1]

Most horse pastures are comprised of cool-season perennial grasses. However, there are times when alternative pasture forages are needed due to extreme weather conditions or horse nutritional needs. Annuals offer an opportunity to provide emergency or short-term forage when weather or other circumstances negatively impact an area. Researchers in Minnesota (MN) explored grazing horses on annual cool- and warm-season grasses. Based on a combination of horse preference, forage yield, and forage quality, annual ryegrass and teff were deemed acceptable cool- and warm-season annuals, respectively, for use in horse pastures. Additionally, some horses may need to graze forages other than cool-season grasses due to metabolic concerns or the need for an increased plane of nutrition. Some horses require a lower amount of nonstructural carbohydrates (NSC) in their diets. Unfortunately, cool-season grasses tend to be higher in NSC compared to legumes and warm-season grasses. In MN, cool-season perennial grasses, alfalfa and teff averaged 11.2, 9.4 and 8.2% NSC (dry matter), respectively, throughout the grazing season. Therefore, grazing teff and alfalfa may be options for horses requiring a lower NSC diet. Additionally, some horses may require more or less digestible energy (DE) based on bodyweight and body condition score. In MN, alfalfa, cool-season perennial grasses and teff averaged 1.08, 0.99 and 0.92 Mcal per pound, respectively, suggesting teff is more suitable for horses with lower energy demands while alfalfa has the capability of meeting requirements for horses with elevated nutritional needs. Grazing alternative forages has a place in horse pasture systems, especially when attempting to alleviate extreme weather conditions or manage certain classes of horses.

[1]Associate Professor, Dept. of Animal Science, University of Minnesota, St. Paul, MN. (krishona@umn.edu).

ECONOMICAL GRAZING MANAGEMENT OF HORSES

Gary Webb[1]

Advantages to allowing horses access to pasture include less cost for harvested forage and stall bedding. In addition, complete or partial daily pasture access will improve attitude and help alleviate muscle soreness in horses which are often ridden on inconsistent schedules. Researchers from North Carolina have measured consumptions of 3% of bodyweight (BW) when horses were grazed for 24 hour, 0.7% in 4 hours or 1.3% BW in 9 hours. These intakes might meet between 10 and 100%, or exceed the requirements of an 1,100 pound horse at light work. With proper management, one acre of pasture will produce from 3 to 7 tons of dry matter (DM) per year. This would provide more than adequate DM for at least one, if not two horses, a year. However, if damaged due to overgrazing during periods of reduced growth, inadequate weed control, low fertility, or drought, forage production per acre could be much less. Because horses have prehensile lips, they have the tendency to overgraze some areas and leave others untouched. Thereby creating roughs and lawns or "spot grazed" areas which will greatly reduce the regrowth of forage in the closely grazed areas. Rotational grazing is a practice which can reduce or eliminate the development of these areas. When choosing fences for rotational grazing cost, ease of fence movement and safety are important. It is possible to purchase and install temporary fencing for rotational grazing for 6 to 12 cents per foot.

[1]Professor, Animal Science Department, Missouri State University, Springfield Missouri. (garywebb@missouristate.edu).

CONTINUOUS GRAZING OF NATIVE WARM-SEASON GRASS IN THE FESCUE BELT

K.A. Brazil, P.D. Keyser, G.E. Bates, and J.T. Mulliniks[1]

Continuous, season-long grazing has traditionally been considered incompatible with native warm-season grasses (NWSG). We determined the productivity and sustainability of continuously grazing a blend of big bluestem (BB), Indiangrass (IG), and little bluestem with beef cattle in the Fescue Belt. Two continuous grazing strategies, continuous (CONT) and heavy-early (HEAVY), were compared in 2015-2017. Using 6 CWT weaned steers, CONT pastures were stocked at a standard rate (1x) throughout the grazing season (105-day average). Pastures assigned as HEAVY were stocked at 1.25x the CONT rate from grazing initiation to 25 June, then reduced to 0.75x the CONT rate for the remainder of the grazing season. Plant population and tiller counts were sampled once a year at grazing initiation. Average daily gains (lb/day) for CONT and HEAVY were similar (2.18 and 1.98, respectively), as was total gain (lb/ac; 342 and 303, respectively). Grazing days (days/ac) did not differ between CONT (156) and HEAVY (152). A potential trend (P = 0.07) in plant population (plants/ft^2) was detected indicating CONT (1.18) may have had greater stand density than HEAVY (1.02), although there was no difference among years. Tiller counts (no/plant) were similar between CONT and HEAVY for BB (86 and 80, respectively) and IG (67 and 67, respectively), and decreased from year one (BB, 106; IG, 105) to year two (BB, 66; IG, 51), and were similar in years two and three (BB 77; IG 45). The decline between years 1 and 2 may have been a result of sampling methodology. Continuous season-long grazing appears to be a viable management option for NWSG in the Fescue Belt.

[1]Graduate Research Asst, University of Tennessee (UT), Knoxville, TN (kbrazil@utk.edu); Professor, UT, Knoxville, Knoxville, TN (pkeyser@utk.edu); Professor, UT, Knoxville, Knoxville, TN (gbates@utk.edu); Asst Professor, University of Nebraska, Lincoln, NE (travis.mulliniks@unl.edu).

CHALLENGES OF ALFALFA PRODUCTION ON WARM-SEASON GRASS PASTURES

L. Rice, K.J. Han, and E. Twidwell[1]

The Gulf coast region needs more productive and persistent perennial legumes to enhance its livestock production efficiency; however, growing alfalfa (*Medicago sativa* L.) in Louisiana has some challenges such as heavy weed competition and soil nutrient deficiency. 'Bulldog 805' alfalfa was planted over warm-season grass pasture in early November. A herbicide study evaluated the effects of Butyrac® (2,4-Dichlorophenoxy butyric acid) on suppression of weed competition. The herbicide was applied at 2 qts/a after early spring harvest. Approximately 95% of broadleaf weeds were controlled and overall crude protein (CP) was increased. Another study tested the effects of potassium fertilization and harvest maturity on DM yield and feed value of the pasture. Potash was applied at the rates of 0, 60, 90, 120, and 150 lbs/acre in the spring, and harvests were made at the bud, early, and mid-flowering stages. Harvest maturity greatly affected both DM yield and feed value of the pasture, but potassium fertilization did not. Bud stage harvest produced pasture of 7% higher CP than the mid-flowering harvest. Neutral detergent fiber (NDF) was 44% in the pasture harvested at the bud stage while it was 50% at the mid-flower stage. Mid-flower stage produced the greatest DM. Based on the first year's data, application of Butyrac® herbicide helped alfalfa compete with weeds until early summer and alfalfa over-seeding contributes to the feed value enhancement of this region's spring pasture.

[1]Forage Agronomy Graduate Assistant, Louisiana State University, Baton Rouge, LA (LRice@agcenter.lsu.edu); Forage Agronomist, Louisiana State University, Baton Rouge, LA (khan@agcenter.lsu.edu); Extension Forage Agronomist, Louisiana State University, Baton Rouge, LA (Etwidwell@agcenter.lsu.edu).

LATE PLANTING OF EARLY VS. LATE MATURING ANNUAL RYEGRASS IN NORTHERN ALABAMA

M. L. Marks[1], M. K. Mullenix, M. K. Stanford, L. A. Kriese-Anderson

Limited information is published on the growth potential of cool-season annual grasses planted outside the recommended window of establishment as an emergency forage option following drought conditions. The objective of this study was to determine the impact of late planting annual ryegrass on seasonal herbage accumulation, grazing season length, and forage nutritive value at the Sand Mountain Research and Extension Center in Crossville, AL. Two annual ryegrass varieties [Winterhawk or Marshall ryegrass] were planted across three planting dates in winter 2016/2017 (December 15; February 1; March 1) into a prepared seedbed. Each variety × planting date combination was replicated in two 2 acre paddocks. Seasonal yield was greatest for Marshall ryegrass planted in March. December planted ryegrass was intermediate in production compared to March planted Marshall, and February planted ryegrass and March planted Winterhawk had the lowest production. Overall ryegrass yield was between 4,000 and 9,000 lb DM/acre across planting dates. Frost events occurred largely in early January, and again in midFebruary. Yield of February ryegrass plantings were more negatively affected by frost events than December or March plantings. Grazing was initiated on December-planted ryegrass in March, whereas February and March dates only were grazed beginning in May. There were 70 days of grazing provided in this trial (March to June 2017).

[1]M. Landon Marks, Regional Extension Agent, Alabama Cooperative Extension System, Centre, AL (mlm0013@aubrun.edu).

GREENHOUSE STUDY OF DIRECT-SEEEDED SWTICHGRASS (*PANCIUM VIRGATUM*) ESTABLISHMENT IN LOWER MISSISSIPPI RIVER SOILS

LaShunda Anderson. Hodges[1], Krishna N. Reddy and Efren Z. Ford[1]

As of October 2016, 32.61% of the Southern U.S. is listed under drought conditions. Drought conditions could adversely impact yields of the traditional crops, such as soybean and cotton. A potential crop in managing soil health during a regional drought is Switchgrass *(Pancium virgatum).* Switchgrass root systems which can grow at least 10 feet deep into the soil are water efficient during drought conditions, increase soil health, as well as, decrease global climate change by increasing carbon sequestration. The overall goal of this greenhouse study was to evaluate the ability of three direct-seeded switchgrass varieties to establish in two Southern U.S. Lower Mississippi River Soils. The three switchgrass varieties are (1) Arkansas Grand Prairie, (2) Cave-In-Rock, and (3) Alamo. The two soils are (1) Tunica Clay (Lowland/Delta) and (2) Memphis Silt Loam (Upland/Loess Bluff). The objectives of this study were (1) to evaluate the ability of three switchgrass varieties to produce direct-seeded seedlings in two Southern U.S. Lower Mississippi River Soils, and (2) to evaluate the effect of soil type upon the establishment of three switchgrass variety direct-seeded transplants. All varieties displayed the ability to establish on Lower Mississippi River soils. Alamo established in Tunica Clay had a significantly different plant height (24.85 in) from all switchgrass varieties established in Tunica and Memphis soils, except for the Arkansas Grand Prairie variety plant height (23.15 in) established in Tunica Clay. Therefore, soil type and switchgrass variety do influence plant height and biomass production in the Lower Mississippi River Region. This indicates that switchgrass could be used as an alternative forage crop to manage soil health during regional drought in the Southern U.S. Lower Mississippi River Soils.

[1]LaShunda Anderson Hodges, llanderson@alcorn.edu, Department of Agriculture, Alcorn State University, Lorman, MS 39096; Krishna N. Reddy, krishna.reddy@ars.usda.gov, USDA-ARS Crop Production Systems Research, Stoneville, MS 38776, Efren Z. Ford, efren.ford@ars.usda.gov. USDA-ARS Crop Production Systems Research, Stoneville, MS 38876.

SCREENING CHEMICAL CONTROL OPTIONS FOR THE BERMUDAGRASS STEM MAGGOT

Lisa Baxter, Holly Anderson, Philip Edwards, Jeremy Kichler, Jenna Kicklighter, Carole Knight, Will Lovett, Peyton Sapp, Matthew Studstill, Joseph Studstill, and Dennis Hancock[1]

Bermudagrass (*Cynodon dactylon*) is one of the primary forage crops grown in the Southeastern US. Unfortunately, the recent invasion of the bermudagrass stem maggot (BSM; *Atherigona reversura* Villenueve) has damaged bermudagrass pastures and hayfields. Strategically-timed pyrethroid applications can significantly reduce adult BSM populations and protect bermudagrass during the most sensitive regrowth phase. Overuse of pesticides of a single mode of action to combat a pest capable of producing a large number of offspring is likely to eventually result in resistance. The objective of this research was to evaluate the effectiveness of four insecticides on suppressing the BSM. Eight County Agents were randomly assigned an insecticide (zeta-cypermethrin, malathion, carbaryl, or spinosad) to be applied on a grower's field. Agents swept the field for BSM before and after each spray application to determine effectiveness of the chemical. The zeta-cypermethrin and malathion most effectively reduced BSM populations ($P < 0.001$; 91% and 84% reduction, respectively), followed by carbaryl (41%) and spinosad (21%). These results will be used to design on-farm experiments for 2018.

[1]Post-doctorate, Crops and Soils Science Dept., University of Georgia, Tifton, GA (baxterl@uga.edu); County Extension Agent, Ben Hill County Extension, Fitzgerald, GA, (hhickey@uga.edu); County Extension Coordinator, Irwin County Extension, Ocilla, GA, (pedwards@uga.edu); County Extension Coordinator, Colquitt County Extension, Moultrie, GA, (jkickler@uga.edu); County Extension Agent, Colquitt County Extension, Moultrie, GA, (jkickli2@uga.edu); County Extension Agent, Bulloch County Extension, Statesboro, GA, (clh@uga.edu); County Extension Coordinator, Bacon County Extension, Alma, GA, (welovett@uga.edu); County Extension Coordinator, Burke County Extension, Waynesboro, GA, (psapp@uga.edu); Georgia cattle producer, Abbeville, GA; Georgia cattle producer, Abbeville, GA; Forage Extension Specialist, Crop and Soils Science Dept., University of Georgia, Athens, GA, (dhancock@uga.edu)

ECONOMIC IMPACT OF THE BERMUDAGRASS STEM MAGGOT

Lisa Baxter, William Anderson, and Dennis Hancock[1]

Since it was discovered in South Georgia in the summer of 2010, the bermudagrass stem maggot (BSM; *Atherigona reversura* Villenueve) has severely damaged bermudagrass (*Cynodon dactylon*) pastures and hayfields throughout the Southeast US. Although the degree of damage depends on the bermudagrass variety, latitude of the farm, and time of year, producers have reported up to an 80% yield loss in late summer. The objective of this research was to quantify the damage potential of the BSM in terms of forage yield and nutritive value on five bermudagrass varieties, and estimate the economic impact of this damage. The severity of damage increased throughout the growing season ($P < 0.01$), exceeding 60% yield loss in the fine-textured varieties such as Alicia or Russell in the late season harvests of the first two years. Preliminary research has shown that BSM damage decreased Relative Feed Quality (RFQ) of late-season bermudagrass hay by 7% on average ($P < 0.01$). This decrease was attributed to lower Total Digestible Nutrients (TDN; $P < 0.01$) and lower dry matter intake (DMI; $P < 0.01$). Crude protein (CP) increased in the damaged bermudagrass ($P < 0.01$), but this was a function of dilution of desirable carbohydrates. The economic impact of the BSM damage depends on several factors, but if a conservative yield loss of 25% is generally applied to just the bermudagrass acreage in Georgia, the potential economic loss totals $30 million.

[1]Post-doctorate, Crops and Soils Science Dept., University of Georgia, Tifton, GA (baxterl@uga.edu); Research Geneticist, USDA-ARS, Tifton, GA (bill.anderson@ars.usda.gov); Forage Extension Specialist, Crop and Soils Science Dept., University of Georgia, Athens, GA, (dhancock@uga.edu)

EFFECTS OF CHOPPING ON FERMENTATIVE AND NUTRITIONAL TRAITS OF RYEGRASS, KIKUYU, AND REED CANARYGRASS SILAGE

L. Villalobos[1] and J.A. Arce[2]

Although it is common practice to chop grasses when making silage, producers are interested in reducing labor costs by making silage without chopping. We evaluated the effects of chopping perennial ryegrass (*Lolium perenne* L.), kikuyu (*Kikuyuocloa clandestina*), and reedcanary grass (*Phalaris arundinacea* L.), at 60, 90, or 70 days of regrowth, respectively. Forage was harvested, wilted for 24 hours, and either chopped or unchopped. Molasses (3%) and a farm-made inoculant (0.1%) were mixed for all four microsilo bags filled (1 kg each, 0.063 mm thick) per treatment (4 replicates x 2 treatments x 3 grass species = 24 microsilos). We extracted air with a vacuum and sealed with duct tape. Dry matter differed among grass species and only kikuyu had lower values from chopping (20.9% vs. 23.6%). Protein did not differ among species or treatments, and fiber was affected by grass species (ryegrass 49.6%, reedcanary grass 51.5%, and kikuyu 62.5%). Ryegrass silage was more digestible (63.3%) than reedcanary grass (54.8%) and kikuyu (47.2%). While pH values were higher than 4 for all silages, ammonia content was highest in unchopped ryegrass (6.4%) and lowest in kikuyu (avg. 1.2%). Buffer capacity did not differ among treatment groups in kikuyu grass (avg. 80.1 mEq NaOH.100 g^{-1} DM), but was lower than in the other two species. Aerobic spoilage was not evident in any of the silos but fermentative traits would likely shorten duration of storage and feed out phases. Differences in nutritional and fermentative traits of silages were attributed to differences among species. We did not find evidence that chopping affected silage quality for the grass species evaluated.

[1]Research Center for Animal Nutrition, Department of Animal Science, University of Costa Rica, San José, Costa Rica (luis.villalobosvillalobos@ucr.ac.cr). [2]Department of Animal Science, University of Florida (jose.arcecordero@ucr.ac.cr)

MANAGEMENT STRATEGIES FOR PASTURE RECOVERY IN GUINEAGRASS (*Panicum maximum* Jacq.) AND STARGRASS (*Cynodon nlemfuensis* Vanderyst) PASTURES

L. Villalobos[1] and R. WingChing-Jones[1]

Degradation of pastures can occur when senescent material accumulates, which in turn limits an effective grazing. We evaluated the effects of reducing the senescent material in African Stargrass (*Cynodon nlemfuensis* Vanderyst) and Guineagrass (*Panicum maximum* Jacq. cv Mombasa) pastures. We cut Stargrass using a grass mower BCS® 630 Max and Guineagrass with a scythe at 5, and 15 cm stubble height, respectively. We hand-clipped samples at 5, and 10 cm height at 25, and 17 days of regrowth for Stargrass and Guineagrass plots. Biomass, structural components, and nutritional value were analyzed in the Research Center for Animal Nutrition at the University of Costa Rica. We found higher biomass yields in the control plots for both grasses, having similar values for Stargrass (4154 vs 4796 kg DM.ha^{-1}) and higher differences were found in Guineagrass (5063 vs 13661 kg DM.ha^{-1}). The treated plots had higher percent of leaves in Stargrass (42.6 vs 34.3 %) and Guineagrass (38.7 vs 68.4 %), resulting in an increase in the leaf-to-stem ratio for both. The changes in the structural components favored higher protein contents in Stargrass (16.5 vs 13.4 %) and Guineagrass (9.9 vs 6.3 %). Fiber content was similar in Stargrass (62.5 vs 63.7 %) but its digestibility was almost five percent higher in the treated plots (58.4 vs 53.7 %). The lower fiber values found in Guineagrass in the treated plots (65.7 vs 72.6 %), increased its digestibility (67.2 vs 47.5 %). Guineagrass was the species with greater changes in nutritional value because of the changes in structural components. The lower yields obtained in the treated plots can be offset by the potential higher dry matter intakes in the cattle grazing pastures with leafier growth after senescent material is reduced.

[1]Research Center for Animal Nutrition, Department of Animal Science, University of Costa Rica, San José, Costa Rica. Corresponding author: luis.villalobosvillalobos@ucr.ac.cr

MIXED LIVESTOCK GRAZING
LATE SUMMER HAY FEEDING TO REST AND STOCKPILE PASTURES

Greg Brann[1]

Mixed Livestock Grazing: Mixed grazing has several benefits: increased utilization of forage, improved manure distribution, improved control of internal parasites, improved predator control and increased meat production per acre. Mixed grazing can be done several ways: all in one group, leader follower grazing or separate groups on different pastures. Some disadvantages of adding small ruminants to a cattle operation include: increased fencing, predator issues, feet issues, susceptibility to internal parasites, a higher quality diet is needed and close grazing preference of sheep. These issues can generally be overcome by selecting good livestock, managing grazing height above 5" and birthing when grass is growing.

Late Summer Hay Feeding: One of the most effective ways to reduce the overall amount of hay fed is feeding hay in late summer and early fall when growing conditions are ideal. This is because pasture availability is typically low in early fall and overgrazing at this time limits pasture growth during this potentially ideal growing period. Grazing managers that maintain minimum grazing heights or higher and rotate stock at least two times a week stand to benefit the most since their grazing utilization is higher.

Boom and Bust Grazing: Grazing practices like "take half, leave half" or "top third grazing" allows animal's selectivity which can lead to undesirable vegetation increasing. Periodic close grazing to increase consumption of undesirable plants can reset the plant community to a vegetative state improving palatability and light utilization.

[1]Grazing Soil Health Specialist and Cattle/Sheep Farmer, TACD, Adolphus, KY (gregbrann5@gmail.com).

BALE GRAZING AND ANNUAL FORAGES IN A COW-CALF SYSTEM

Beth Hodges[1]

Bale Grazing: Traditional winter hay feeding in the Upper South region involves using a tractor multiple times per week and at the same time concentrates waste nutrients in small areas. Bale grazing is a winter hay feeding technique where all feeding operations can be completed by hand using temporary electric fencing. Bales are spread out in a checkerboard fashion and thus nutrients are distributed evenly across targeted pastures. The combined result is a tremendous savings in machinery costs and eliminates the need for commercial fertilizer for pasture and potentially hay ground. Secondary benefits are the potential for improved nutrition while feeding hay and improved animal health.

Annual Forages in a Cow-Calf System: Millions of acres were taken out of cattle production and put into grain production during the ethanol boom years of 2008-2013. With the significant drop in grain prices since 2014, much of this marginal grain ground will go back into livestock production. Rather than seeding back into a typical (and often expensive) perennial mixture, this ground can be seeded into a low-cost summer and winter annual forage system. An advantage of using an annual forage system is that it allows either an increase in stocking rate or a decrease in hay feeding. Many cattle farmers do not have ready access to a drill. The system detailed in this presentation uses forages that can be broadcast successfully.

[1]Cattle and Grain Farmer, Double B Farms, Summersville, Kentucky
(beth@hodgesventuresllc.com).

VIRGINIA TECH'S SUMMER STOCKPILE SYSTEM: DEFERRED SPRING GROWTH FOR SUMMER GRAZING

Chris Teutsch[1]

Grazing systems in the Upper South suffer from forage deficits during both the summer and winter months. Research conducted at Virginia Tech's Southern Piedmont and Shenandoah Valley Agricultural Research and Extension Centers evaluated deferred spring and early summer tall fescue growth to provide intensive grazing during the summer and early fall. This system could be a relatively simple and economical way to provide needed forage during the summer months. In addition, setting aside fescue pastures in August for fall stockpiling in the Upper South has been researched and promoted for decades. However, an inconvenient problem often ignored is that we are typically low on pasture availability at this time of year, and it is difficult to set aside pasture for traditional stockpiling. Utilizing accumulated pasture growth from spring and early summer to extend grazing during late summer and early fall could provide needed forage while other pastures within the grazing system are rested and growth is accumulated for late fall and winter grazing. Key concepts and questions covered in this presentation include:

- How the overall combined summer/fall stockpile system works
- How many extra grazing days can be expected from this system
- Nutritional quality and animal performance in these systems
- The use of both KY-31 and novel endophyte tall fescue in the system

[1]Associate Extension Professor, Plant and Soil Science, University of Kentucky
(chris.teutsch@uky.edu)

EVALUATION OF FIVE WARM-SEASON FORAGES FOR BEEF CATTLE PRODUCTION

K.E. Zechiel, J.D. Rhinehart, P.D. Keyser, G.E. Bates, A.J. Ashworth, and C.N. Boyer[1]

Increased frequency and severity of droughts years have emphasized the importance of reliable, drought-resilient summer forages for fescue-belt producers. We compared five alternatives at two locations (Springfield and Grand Junction, TN) over three years (2014-2016) for weaned heifers (521 lb initial weight): an improved bermudagrass (BG), eastern gamagrass (EG), lowland switchgrass (SW), a big bluestem and indiangrass blend (BBI), and crabgrass (CG). Nitrogen fertilization rates (60 Units N) were set to align with typical application during drought conditions rather than per species BMPs. Average daily gain (lb/day) was greatest for BBI (1.36) and least for BG (0.90), but did not differ among CG, EG, and SW (0.97, 0.92, and 1.12, respectively). Grazing days/ac were greatest for EG (269) and SW (250) and similar for BBI, BG, and CG (167, 186, and 184, respectively). Total gain (lb/ac) was greatest for SW (280) and EG (246) and lowest for BG (166) and CG (178); BBI was intermediate at 231. Crude protein ranged from 9.4-11.5% but did not differ among species. Neutral detergent fiber was lowest for CG (60.1%) and BG (61.5%) and greatest for SW (68.0%) and EG (67.9%); BBI was intermediate at 66.4%. Economic analyses indicated that except for CG (planted annually), all per acre net returns were positive with SW ($147), EG ($110), and BBI ($96) all exceeding BG ($40). Strengths of each summer forage option (e.g., rates of gain, carrying capacity, response to N) should be matched to producer objectives to select which best meets summer forage needs.

[1]Graduate Research Asst, University of Tennessee, Knoxville, TN, (kzechiel@vols.utk.edu); Assoc Professor, UT, Spring Hill, TN (jrhinehart@utk.edu); Professor, UT, Knoxville, TN (pkeyser@utk.edu); Professor, UT, Knoxville, TN (gbates@utk.edu); Research Scientist, USDA-ARS, Fayetteville, AR (Amanda.Ashworth@ARS.USDA.GOV); Assoc Professor, UT, Knoxville, TN (cboyer3@utk.edu).

COVER CROP ECONOMICS

Chris Prevatt, Jennifer Johnson Tucker, Kim Mullenix, Mace Bauer,
Deidre Harmon, and Marcelo Wallau

Many soils throughout the United States have been eroded and degraded from continuous crop production or overgrazing from livestock. Despite the widespread recognition of cover crops as a means for increasing soil health by reducing soil erosion and nutrient loss, many producers have yet to adopt cover crops. One common perspective from producers is that while cover crops are good for the soil, they are not a cost-effective production practice for their operation. While long-term benefits to the land from planting and grazing cover crops are hard to determine, we can easily calculate the short-run costs and benefits for producers who chose to graze their cover crops. An excel spreadsheet was developed to compare and analyze the potential costs and returns for planting and grazing various winter and summer annual cover crops. This tool can help producers determine if cover crops are an economically viable production practice for their operation. Cover crop plants that were used in this analysis included grasses, legumes, broadleaves, and brassicas. Cover crop yields and animals gains were analyzed to help producers determine their approximate breakeven point. The excel spreadsheet used in this analysis is available here: http://rcrec-ona.ifas.ufl.edu/Economics/index.shtml. This poster looks to raise awareness of the potential benefits and costs for farmers adopting cover crops. We'll share the information we've learn about cover crops and propose how to put this information into practice.

AN ECONOMIC ANALYSIS OF INTERSEEDING COOL-SEASON LEGUMES INTO PERENNIAL GRASS PASTURES

Chris Prevatt, Kim Mullenix, Katie Mason, Jennifer Johnson Tucker

While perennial grass pastures provide the basis for beef production systems, interseeding cool-season legumes into these pastures can serve as complementary resources for filling in forage production gaps, reducing stored feed needs, reducing the need for nitrogen fertilizer, increasing the nutritional quality of forage, increasing biomass production, improving animal performance, and reducing the toxic effects from endophyte-infected fall fescue. An economic analysis was conducted to measure the potential benefits and added costs of interseeding cool-season legumes into grazed perennial grass pastures. An excel spreadsheet was developed as a tool to help producers determine if interseeding cool-season legumes into their perennial grass pastures can be an economically viable production practice for their operation. The economic factors included in the spreadsheet were a) cost of cool-season legume seed, b) cost of seeding application, c) increase in average daily gain, d) increase in forage production, e) grazing utilization, and g) cost per dry matter ton. The total cost, benefit, and net return of interseeding cool-season legumes into perennial grass pastures were calculated by the excel spreadsheet. The excel spreadsheet used in this analysis is available here: http://rcrec-ona.ifas.ufl.edu/Economics/index.shtml.

FORAGE TESTING FOR LIVESTOCK NUTRITION AND CROP NUTRIENT MANAGEMENT

E. B. Rayburn[1]

Forage testing is the foundation for cost effective livestock feeding and crop nutrient management. Beef producers in West Virginia have found that they can save $10 for every $1 invested in forage testing when using the test results to fine tune the supplements they feed their livestock. The question often asked is, how accurate is a forage test based on the small sample sent to the laboratory? To answer this question we sampled hay from 40 hay fields. From each field we took two forage samples, for a total of 80 forage samples. Each of the two samples were taken from 8 to 12 large round hay bales (800 lbs. weight) harvested from a given field, on the same day, managed in the same way. Each sample filled a 1 quart zip-lock plastic bag. The two samples represented 16 to 24 bales per field. To measure how accurate a sample representing 8 to 12 bales is for determining the average nutrient content of the hay, the nutrient content of individual samples was compared to the average of the two samples by regression analysis. Based on this analysis, two thirds of the time crude protein (CP), total digestible nutrients (TDN), and phosphorus (P) for a sample represent 8 to 12 bale was within 3 to 4% of the sample representing 16 to 24 bales. For the minerals calcium (Ca), magnesium (Mg), and zinc (Zn) the single analysis was within 8 to 11% of the average of the two samples, two thirds of the time. This indicates that taking a forage sample from as few as 8 to 12 bales from a uniformly managed hay field is adequate for evaluating the hay for beef cattle nutrition and plant nutrients removed by the hay crop.

[1]Extension Specialist, West Virginia University, Morgantown, WV (erayburn@wvu.edu).

IMPACT OF MICROBIAL INOCULANTS IN THE FERMENTATIVE PROCESS OF BAHIAGRASS/ALFALFA HAYLAGE

R. Lemus and J.A White[1]

The increasing popularity of interseeding alfalfa into perennial warm-season grasses such as bermudagrass and bahiagrass has created a need for more research involving this system in the southern USA. A study was conducted to measure the efficacy of microbial inoculants in improving the fermentation and nutritive value of bahiagrass/alfalfa baleage on a farm scale. The study was conducted in New Hebron, MS on a 25-ac field of 'Argentine' bahiagrass. 'Bulldog 805' alfalfa was inter-seeded at a rate of 20 lb PLS/ac. Treatments included a control and five inoculant treatments (Pioneer 11H50, Pioneer 1174, Pioneer 11G22, SiloSolve MC, and SiloKing) applied at the manufacture recommended rates. Treatments were applied during the second and third harvests in 2016 when there was a more equal ratio of alfalfa and bahiagrass. Six large round bales (4'x5') were treated with each inoculant and wrapped with 8 layers of UV-treated plastic. Temperature micro-sensors were inserted in to the bale and data collected for 21 d. Forage samples were taken 18 and 24 inches deep at 0, 7, 14, 21 and 112 d after ensiling. Samples were analyzed for pH and nutritive value. The average DM at each harvest was 31.3 and 56.4%. There was an inoculant treatment effect for pH and CP during the harvest 1 (H1) and 2 (H2). Data indicated a stable pH at 7 d post-ensiling along with a temperature equilibration. The pH was lower in H2 and CP was higher for all treatments in H1 compared to H2.

[1]Extension Forage Specialist, Center for Forage Management & Environmental Stewardship, Mississippi State University (Rocky.Lemus@msstate.edu); Forage Variety Testing Manager, Mississippi Agricultural & Forestry Experiment Station, Mississippi State University (Joshua.White@msstate.edu).

ANNUAL RYEGRASS BIOMASS PRODUCTION FOLLOWING THE APPLICATION OF BIOWASH100

R. Lemus and J.A White[1]

Fertility deficiency, especially nitrogen, affects forage production on a large scale and is one of the most common nutritional problems among livestock producers. A host of products commonly known as additives or enhancers typically include stabilizers, adjuvants, or plant growth regulators has been marketed to enhance nitrogen activity. A two-year study was conducted to examine the effects of BioWash 100 on biomass production, nutrient uptake and nutritive value of 'Marshal' annual ryegrass. The study was conducted in Starkville, Mississippi. The experimental design was a randomized complete block with six treatments replicated four times. The treatments included a control, BioWash100 applied at two rates (5 and 10 oz/ac), 32% AUN (50 lb N/ac), and AUN plus BioWash 100 rates. Annual ryegrass was planted with a drill at a rate of 20 lb/ac in a prepared seedbed. First treatment application occurred after germination when plants have reached 3 inches of growth and each subsequent treatment was applied three days after harvest. Nitrogen was applied not to exceed 150 lb N/ac/yr. Treatments were harvested when at least 50% of the plots had reached 12 to 15 inches of growth. Forage yields of plots receiving Biowash 100 only, were similar to the control at both rates of application. Adding Biowash 100 to the liquid urea ammonium nitrate did not provide a significant increase in seasonal or harvest yield. Very little effect was observed in the nutritive value of annual ryegrass with BioWash100 application.

[1]Extension Forage Specialist, Center for Forage Management & Environmental Stewardship, Mississippi State University (Rocky.Lemus@msstate.edu); Forage Variety Testing Manager, Mississippi Agricultural & Forestry Experiment Station, Mississippi State University (Joshua.White@msstate.edu).

DEVELOPING RENOVATION STRATEGIES FOR TOXIC-INFECTED TALL FESCUE THAT BEST MANAGES FOR PROFITABILITY, ANIMAL PERFORMANCE AND SOIL HEALTH OF THE PRODUCTION SYSTEM.

S.H. Ingram, M.H. Poore, A.J. Franzluebbers, M.S. Castillo, and D.H. Poole[1]

Kentucky 31 tall fescue (TF) is the most widely cultivated forage in the U.S. After rapid adoption, farmers noticed poor growth performance, reproductive issues and overall poor body condition of animals grazing this forage. The reason for these symptoms were ergot alkaloids produced by a fungal endophyte within the fescue plant. The best option for improving plant and animal performance is achieved by planting a novel endophyte TF variety (NE). The intent of this study is to determine effectiveness of three renovation strategies by analyzing profitability, animal performance, agronomic performance, soil heath, and presence of toxic infected tall fescue after establishment. Planting NE will take place in the fall of year 3 for all three conversion strategies. Treatments evaluated are, 1) a control, 2) conversion to NE after a single season, single specie cover crop 3) conversion to NE after three seasons of a single specie cover crop, and 4) conversion to NE after three seasons of a multi-specie cover crop. Baseline soil measurements to estimate soil health did not differ among treatments. Agronomic performance of cover crops and animal performance is expected to be greater for all conversion strategies in comparison to the control. It is expected for treatment 3 and 4 to recoup more of the cost of conversion and decrease presence of toxic infected tall fescue in comparison to treatment 2.

[1]Research Assistant, North Carolina State University, Raleigh, NC (shingram@ncsu.edu); Beef Cattle Specialist, North Carolina State University, Raleigh, NC (mhpoore@ncsu.edu); Soil Ecologist, USDA-ARS, North Carolina State University, Raleigh, NC (ajfranzl@ncsu.edu); Forage Specialist, North Carolina State University, Raleigh, NC (mscastil@ncsu.edu); Reproductive Physiologist, North Carolina State University, Raleigh, NC (dhpoole@ncsu.edu).

EFFECTS ON SCROTAL CIRCUMFERENCE AND BSE OUTCOME ON BULLS IN AN ALL FORAGE BULL TEST VS. A FORAGE BULL TEST WITH SUPPLEMENTATION

G.S. Sell[1], S.L. Pratt[2], J.G. Andrae[3]

Since 1983 Clemson University has operated a forage bull test station at its Edisto Research and Education Center located near Blackville, SC. From 1983 until 2010 the test was conducted as a forage bull test with supplementation. By definition this means the bulls were grazed on forages and not in a feedlot environment, but were supplemented daily during the testing period of 168 days with a ration fed up to 1% of their body weight. From 2011 to present the bulls have been tested in a truly forage only bull test. They are not supplemented with any grain during the 168 day testing period and are tested solely on forages. This presented us with a unique opportunity for an analysis comparing the two management styles. A large body of data was analyzed and we found that there are significant differences in growth traits between an all forage test and a supplemented forage test, including ADG, WDA, and Sale Index, with bulls in a forage test with supplementation having a higher mean for all these traits ($P<0.05$). We also found that while test management style did have a significant effect on scrotal circumference at day 112 ($P<0.05$), this effect was removed by the time of the BSE at day 332 ($P<0.05$). Management style had no effect on SC at the BSE, nor did it have an effect on the pass/fail rate of the BSE ($P<0.05$). When a linear regression was run for body weight by age for bulls on an all forage diet vs. a forage diet with supplementation, we found there may be sufficient evidence to warrant adjusting 365d and 452d adjusted yearling weight formulas for bulls on an all forage test. Further analysis is needed however.

[1] Extension/Research Associate, Clemson University, Edisto Research and Education Center, Blackville, SC (gsell@clemson.edu); [2]Professor, Clemson University, Animal and Veterinary Science Dep., Clemson, SC (scottp@clemson.edu); [3]Professor/Extension Forage Specialist/REC Director, Clemson University, Simpson Research and Education Center, Clemson, SC. (jandrae@clemson.edu).

HAY QUALITY LOSS DURING STORAGE

G.E. Shewmaker[1]

The effects of several months storage in 4 stacks of hay was studied to quantify changes in forage quality parameters. The same large rectangular bales in each stack were sampled by compositing 20 cores soon after stacking and after 6 months of storage near Kimberly, ID. Crude protein in one stack increased 2.1units (% of dry matter) from initial to final storage, but the three other stacks did not change for crude protein. The average across four stacks for acid detergent fiber (ADF) increased from 2.7 to 5.3 units (% if dry matter) from initial to final sampling. The change in neutral detergent fiber (aNDF) was inconsistent for the four hays. The RFV index did not change in the alfalfa/grass mix hay, but declined by 14, 21, and 11 units in the other three alfalfa hays. The NDF digestibility in 48 hours (NDFD48, as % of NDF) declined an average of 2.9 units (% of dry matter) across the four hays. The net energy for lactation (NE/Lact), digestible dry matter, relative feed value (RFV), and relative forage quality (RFQ) declined about the same magnitude across the four hays. Lignin, an indigestible fiber fraction, increased an average of 4.3% across the four hays. This paper documents the changes of hay quality in large rectangular hay bales of <10% moisture in the stack for several months storage from harvest to winter near Kimberly, ID.

[1]Extension Forage Specialist and Professor, University of Idaho, Kimberly Research and Extension Center, Kimberly, ID; (gshew@uidaho.edu).

COMPARISON OF FALL VS. SPRING PLANTING FOR DM YIELD OF WINTER ANNUAL FORAGES

K. J. Simon, J. A. Jennings, and D. Philipp[1]

Every year, producers start asking in February what to plant for emergency spring forage. The typical answer would be some species of winter annual forage. However, yield differences and growth patterns for fall vs. late winter planting of various winter annual forages are not well documented. Seven winter annual forage species were planted in October and March at the University of Arkansas Watershed Research and Extension Center in Fayetteville to evaluate the influence of planting date on spring dry matter production. The winter annual forage species included: Elbon rye, AGS 2027 winter wheat, Jerry spring oat, Coker 227 winter oat, Dixie crimson clover, WinterHawk annual ryegrass, and Winfred forage brassica. Fall-planted winter annuals produced forage earlier in spring than spring-planted annuals. Fall-planted winter annuals were harvested in mid-April. DM yield was over 3,400 lbs/ac for all the treatments except Jerry spring oat which had 100% winter kill. Elbon rye produced the most DM yield at 8600 lbs/ac. Most spring-planted winter annuals reached the harvestable stage about 1 month later than for fall-planted annuals. Spring-planted winter annuals were harvested May 25. DM yield ranged from 5,000 to 550 lbs/ac. Jerry oat produced the most DM yield. Yield of spring-planted wheat, rye, clover, and forage brassica was about half or less compared to the fall-planting. The best potential yield for spring-plantings appears to be winter or spring oat and annual ryegrass.

[1]Animal Science Program Associate – Forages (ksimon@uaex.edu), Professor and Extension Forage Specialist (jjennings@uaex.edu), and Associate Professor-Forages (dphilipp@uark.edu), University of Arkansas Division of Agriculture. Little Rock, Arkansas. 72204

CURRENT TECHNOLOGY AND ADVANCEMENTS IN FORAGE HARVESTING

Kevin Smith[1]

There is currently a wide range of equipment available to transform a standing forage crop in the field to a packaged bundle ready for livestock consumption. Whether the end goal is hay, baleage or silage there are various features available which allow the producer to complete the task more efficiently and with better quality. This presentation will provide a summary of the latest advancements in forage harvesting equipment, from mowers and swath manipulation, to baling and forage harvesters. It will also include some best practice set up and operating techniques to help the producer achieve the results he wants.

[1]Project Engineer, CNH Industrial.

DETERMINING THE EFFECT OF PLANTING DATE AND LAND PREPARATION METHOD ON FORAGE YIELD AND FORAGE QUALITY OF A FORAGE TYPE BRASSICA WHEN GROWN IN GEORGIA

Tayler Denman[1] and Dennis Hancock[2]

Forage brassicas (*Brassica spp.*) are gaining in popularity as a high quality winter forage in the Southeastern USA, yet research that defines best management practices for establishment is limited. Our objective was to examine the effects of planting date and land preparation method on forage yield and quality of a forage-type, hybrid brassica (*B. rapa* subsp. *rapa* x *B. napus*). In 2016, four planting methods (conventional till, no-till after burning, no-till after mowing, and no-till with no residue removal) were compared across four planting dates (1 and 15 September, 1 and 15 October). Forage samples were clipped at 30, 45, 60 and 90 days after planting (DAP) to determine yield. Forage quality was measured at 60 and 90 DAP. There were significant effects of forage yield by planting date, land preparation method, and an interaction of the two factors. For the 1 September treatments at 90 DAP, forage yields in the conventional till were higher than the no-till/burn, and both were higher than no-till/mow or no-till/residue treatments, which were not different from each other (4670, 2150, 116, vs. 0 kg DM ha^{-1} respectively). In planting at 15 October, the yields at 90 DAP showed no difference in conventional till and no-till/burn treatments, but both were higher than the no-till/mow or no-till/residue treatments, which again were not different from each other (1610, 1320, 80, vs. 0 kg DM ha^{-1} respectively). Nutritive value of brassica was exceptionally high, and there was no difference in forage quality across treatments. Forage brassica should be planted early September for highest yields. Regardless of planting date, brassica should be planted into little to no residue for best results.

[1]Graduate Student, University of Georgia, Athens GA, [2]Forage Extension Specialist, University of Georgia, Athens GA.

EVALUATING YIELD, NUTRITIVE VALUE, AND STORAGE LENGTH OF TIFTON-85 BERMUDAGRASS AND TIFTON-85 BERMUDAGRASS-ALFALFA MIXTURES AS BALEAGE IN THE SOUTHEAST.

T.J. Hendricks, J.J. Tucker, D.W. Hancock, R.L. Stewart, Jr.[1]

Interseeding a legume into bermudagrass can improve relative forage quality (RFQ) and reduce N fertilization. Our objective was to compare nutritive value and yield of bermudagrass with and without alfalfa interseeded. Ten 0.5-A 'Tifton 85' (T85) bermudagrass plots were randomly assigned to either T85 or T85 interseeded with 'Bulldog 805' alfalfa on 19 February 2016. T85 bermudagrass-only received 75 lbs N/acre four times throughout the growing season. Plots were harvested at early bloom stage and every 28 to 35 days thereafter, baled at 40-60% moisture, and individually wrapped. Plots were evaluated at each harvest for botanical composition and forage yield. Bales were sampled prior to wrapping and at 6-week, 9 and 12-month time points for nutritive value. In year 1, bermudagrass-only yield was higher than alfalfa-bermudagrass (10,569 vs. 7475 lbs/acre, respectively). However, in year 2, alfalfa-bermudagrass had higher yield than bermudagrass-only (11,201 vs. 6672 lbs/acre, respectively). Cumulative yield over the study were greater in the alfalfa-bermudagrass treatment (18,676 vs. 17,240 lbs/acre, respectively). Year one crude protein (CP) and in-vitro true digestibility (IVTD) were higher in alfalfa-bermudagrass than bermudagrass (14.0 vs. 11.4% CP and 66.7 vs. 60.2% IVTD, respectively). Additionally, the neutral and acid detergent fibers (NDF and ADF) were lower in alfalfa-bermudagrass treatments when compared to bermudagrass-only treatments (55.7 vs. 68.1% NDF and 28.7 vs. 31.9% ADF, respectively).

[1] Graduate Research Assistant, Dept. of Animal and Dairy Science (thend10@uga.edu); Assistant Professor, Dept. of Animal and Diary Science (jjtucker@uga.edu); Associate Professor, Dept. of Crop and Soils (dhancock@uga.edu); Associate Professor, Dept. of Animal and Dairy Science (lawtons@uga.edu), University of Georgia.

EVALUATING THE EFFECTIVENESS OF FIVE COMMERCIALLY AVAILABLE MICROBIAL INOCULANTS FOR IMPROVED FERMENTATION AND AEROBIC STABILITY IN ALFALFA-BERMUDAGRASS BALEAGE IN THE SOUTHEAST

T.J. Hendricks, D.W. Hancock, J.J. Tucker, R.L. Stewart, Jr.[1]

Microbial inoculants are widely used for preservation of hay and baleage. Our objective was to determine the efficacy of five commercial inoculants for improving fermentation characteristics of alfalfa-bermudagrass harvested as baleage. This study was conducted on established Tifton-44 bermudagrass and Bulldog 505 alfalfa at the J. Phil Campbell Research and Extension Center. Fields were harvested 3 May 2017 and 19 July 2017 at 10% bloom stage at a moisture of 55-62%. Bales were randomly assigned to one of 6 treatments: 1) Pioneer 1174 at 11 ppm; 2) Pioneer 11H50 at 11 ppm; 3) Pioneer 11G22 at 11 ppm; 4) SiloKing at 7.9 ppm; 5) SiloSolve MC at 1.1 ppm; and 6) untreated water control. Pioneer and SiloSolve MC inoculants were mixed with 1-gal deionized water and sprayed using a tractor-mounted tank and sprayer, while SiloKing was applied using a dry blower. Both sprayer and blower applicators were oriented to spray the entire windrow before forage entered the baling chamber. Forage was baled to a mean dry-matter density 9.4 lb/ft^3 (7.74 to 15.7 lb/ft^3). Bales were weighed and core samples taken prior to wrapping (pre-storage) at 7, 14, 21, and 60-days post-harvest and on 10 October 2017 (post-storage). Temperature was recorded in two bales per treatment every 30 minutes for 70-days post-harvest. At days 7 and 21 post-harvest, inoculant treatment did not significantly affect bale temperature. However, at 14-days post-harvest, the untreated water control had the lowest bale temperature of any treatment, but all elevated temperatures were not practically significant.

[1] Graduate Research Assistant, Dept. of Animal and Dairy Science (thend10@uga.edu); Associate Professor, Dept. of Crop and Soils (dhancock@uga.edu); Assistant Professor, Dept. of Animal and Diary Science (jjtucker@uga.edu); Associate Professor, Dept. of Animal and Dairy Science (lawtons@uga.edu), University of Georgia.

HIGH TECH ROTATIONAL GRAZING: ADVANCES IN FENCING AND GRAZING MANAGEMENT TOOLS

Scott Flynn[1]

Successful rotational grazing operations monitor forage availability and control animal movement to ensure adequate forage regrowth and to optimize forage quality throughout the season. This allows managers to improve both animal and forage production over the grazing season in comparison to continuous grazing. However, as the intensity of management increases so does labor, requiring managers to find a rotation interval that fits the available labor resources rather than adjusting the interval to maximize production. As technology advances grazing managers are being given more tools to intensify their grazing operations with far less labor than previously required. Phone apps and computer software that allow you to remotely open gates, check water resources, and measure forage production are just a few examples of technology available today. In addition emerging technology will pave the way for virtual fencing, and monitoring grazing behavior and herd health remotely. Leveraging this technology will not only give producers the ability to improve grazing practices and reduce labor, it will also allow operations to be managed over a larger geographic area.

[1]Field Scientist, DowDuPont, Lees Summit, MO (flynn@dow.com).

USING GPS TO MONITOR CATTLE SELECTION OF TALL FESCUE INFECTED WITH NOVEL ENDOPHYTES

Sarah L. Kenyon and Craig A. Roberts[1]

Tall fescue [*Lolium arundinaceum* (Schreb.) Darbysh.] is naturally infected with a fungal endophyte [*Epichloë coenophiala* (Morgan-Jones & W. Gams) C.W. Bacon & Schardl, comb. nov.] that produces toxic ergot alkaloids and cause fescue toxicosis in livestock. Tall fescue has been re-infected with fungal strains called "novel endophytes," which are nontoxic strains of the fungus. The objective was to determine differences in animal preference and stand persistence among commercial novel endophyte-infected tall fescue cultivars. Eight cultivars were planted in 30 x 40 ft plots at Linneus, MO and grazed by non-lactating beef cattle fitted with Loteck GPS collars. Plots were grazed between March and November five times in 2013 and four times in 2014. Grazing began when forage reached 2200 – 2800 lb/acre and concluded when one plot reached 3 inches height. Latitude and longitude were recorded every 5 min to indicate cattle location. Altitude and angle of head were also recorded every 5 min to indicate active grazing. Forage samples were collected from each plot and analyzed for crude protein, acid detergent fiber (ADF), and neutral detergent fiber (NDF). 'BarOptima' tall fescue was grazed first and more frequently than other tall fescue cultivars, whether they were endophyte-free or infected with a novel or toxic endophyte. Cultivars did not differ in crude protein, ADF or NDF, which was expected as forage was maintained in a vegetative state. Further differences in cultivar selection may have been detected with a grazing strategy that included long rotations or continuous defoliation. In addition, rust (*Puccinia* spp.) susceptibility differed among cultivars.

[1]Regional Agronomy Specialist, University of Missouri Extension, West Plains, MO (KenyonS@missouri.edu); Division of Plant Sciences, University of Missouri, Columbia, MO (RobertsCr@missouri.edu).

YIELD MONITORING, MAPPING AND VARIABLE RATE FERTILIZATION FOR HAY AND SILAGE PRODUCTION

K.R. Kirk, P.J. Loftis, J.G. Andrae, G.S. Sell[1]

Precision agriculture technologies associated with row crop production can be equally beneficial to forage crops. One key aspect of precision agriculture allows the producer to assess and manage for spatial variability in his fields; simply put, different areas of each field have different needs. While most growers generally know where high and low yielding areas of a field are, a yield map can quantify and delineate this understanding. Unfortunately, no yield monitors have been developed or are commercially available for forage crops. We have recently developed and field tested a windrow monitor based on ultrasound technology which allows forage yield mapping. One important application of this knowledge for forage production would be in the form of variable rate nutrient application. Knowledge from the yield map of the amounts of nutrients being removed from the field can be applied to return nutrients for the subsequent harvests; the yield map can provide guidance on how to subdivide the field into separate management zones, each receiving different nutrient rates according to need. Uniform nutrient applications "to meet the average needs" result in under-application and therefore yield deficits in the most productive areas of the field and over-application and waste in the least productive areas. The goal of variable rate nutrient application is not to equalize yields across the field, but to place nutrients at appropriate rates according to need. Variable rate nitrogen applications have been tested and will be discussed along with yield monitoring implications.

[1]Precision Agriculture Engineer, Clemson Univ., Edisto REC, Blackville, SC (kirk2@clemson.edu); Graduate Student, Clemson Univ., Plant & Environmental Sciences, Clemson, SC (ploftis@g.clemson.edu); Director and Forage Specialist, Clemson Univ., Simpson REC, Clemson, SC (jandrae@clemson.edu); Forage Bull Test Coordinator, Clemson Univ., Edisto REC, Blackville, SC (gsell@clemson.edu).

TOLERANCE OF FORAGE SORGHUM CULTIVARS TO SUGARCANE APHID

C.D. Teutsch, R.T. Villanueva, Y. Gonzalez, G.L Olson, and S.R. Smith[1]

Forage sorghum (*Sorghum bicolor* (L.) Moench) could provide a drought tolerant alternative to corn (*Zea mays*, L.) for silage production in the upper South. However, a new pest of sorghum, the sugarcane aphid (*Melanaphis sacchari* (Zehntner) (Hemiptera: Aphididae), could restrict its use. The objective of this study was to document of tolerance of commercially available (12) and experimental (6) varieties of forage sorghum to the sugarcane aphid. In May 2017, cultivars were planted at two locations Kentucky. In late August (Princeton, KY) and late September (Lexington, KY), plots were rated for sugarcane aphid damage. Plots were harvest for yield in late September and October for the Lexington and Princeton locations, respectively. Sugarcane aphid damage ratings ranged from 1.75 to 8.75 and 4.25 to 8.25, for Lexington and Princeton, respectively, and differed between cultivars within each location ($P < 0.05$). Adjusted silage yields ranged from 13.7 to 26.5 and 5.2 to 21.3 ton/A, for Lexington and Princeton, respectively, and differed between cultivars within each location ($P < 0.05$). A linear relationship between sugarcane aphid damage and adjusted silage yield was found at the Princeton location ($P < 0.001$, $r^2 = 0.56$), but not the Lexington location. These initial results indicate that differences in tolerance to sugarcane aphid do exist between forage sorghum cultivars and may provide a simple and cost effective approach to managing sugarcane aphids in the sorghum species.

[1]Associate Extension Professor, Dept. of Plant and Soil Sciences, University of Kentucky, Lexington (chris.teutsch@uky.edu); Assistant Extension Professor, Dept. of Entomology, University of Kentucky, Lexington; Technical/Paraprofessional, Dept. of Entomology, University of Kentucky, Lexington; Agriculture Research Specialist Sr, Dept. of Plant and Soil Sciences, University of Kentucky, Lexington; Extension Professor, Dept. of Plant and Soil Sciences, University of Kentucky, Lexington.

MEASURE, MONITOR, AND MANAGE GRAZING SYSTEMS USING *PADDOCKTRAC*

RL Kallenbach, TR Lock, SA Hamilton, and JA Lory[1]

Grazing management optimizes resources in forage/livestock systems. Farmers who frequently measure and monitor forage growth manage pasture resources appropriately. The monitoring of pasture resources helps producers know when plants achieve ideal grazing and residual heights. This in turn optimizes pasture regrowth, plant biodiversity, livestock production, and ecosystem services. Since 2006, MU's "Grazing Wedge" website helped producers manage forage growth. However, many producers found the weekly measurements of pastures time-consuming. Others report that translating these weekly data into a format that facilitates management is beyond their capacity. The inability to provide farmers a simple technology that collects and processes forage estimates limits data-driven grazing management in Missouri, and nationally. Over the past decade, MU developed the principle components for measuring, collecting and processing forage data from multiple paddocks to optimize grazing management. A sensor-based, automated forage management system is now possible. Our goal is to provide farmers an integrated system that comprises an all-terrain vehicle equipped with a bluetooth-enabled ultrasonic sensor to measure forage height and a smartphone app to deliver the sensor data to MU's "Grazing Wedge" website. This data acquisition and interpretation system transforms farmer understanding of forage growth and allocation to livestock.

[1]Assistant Dean, Agriculture and Natural Resources Extension and Professor, Div. of Plant Sciences, Univ. of Missouri, Columbia (kallenbachr@missouri.edu); Sr. Research Specialist, Div. of Plant Sciences, Univ. of Missouri, Columbia (lockt@missouri.edu); Dairy Specialist, Div. of Animal Sciences, Univ. of Missouri, Columbia (hamiltonsa@missouri.edu); Extension Associate Professor, Div. of Plant Sciences, Univ. of Missouri, Columbia (loryj@missouri.edu).

TWELVE YEARS OF HORSE PASTURE EVALUATION

Krista Lea, Tom C. Keene and S. Ray Smith[1]

The University of Kentucky Horse Pasture Evaluation Program begin in 2005 in response to the Mare Reproductive Loss Syndrome (MRLS) of 2001 with a mission of educating farm owners and managers to improve pasture management. The program provides detailed pasture assessment and recommendations to large and small horse farms. To date, over 200 evaluations have been performed in 21 counties of Kentucky, representing nearly 40,000 farm acres. Small farms are often overgrazed and under fertilized, leading to high percentages of bare soil and weeds in pastures. Our recommendations have encouraged these issues to be addressed with over-seeding, rotational grazing, weed control, and regular soil testing. Large breeding farms often underestimate the amount of toxic tall fescue consumed by their pregnant mares. Botanical composition and laboratory analysis of tall fescue gives these farms the information they need to make informed decisions on fescue management. Some focus on dilution while other choose total eradication of tall fescue. This program has generated a tremendous dataset and yielded both qualitative and quantitative information. Additionally, it has led to meaningful contacts within the equine industry and provided hands-on learning to the undergraduates who provide the labor force each summer. Unintended consequences of pasture recommendations have also been observed, including the spread of warm season grassy weeds into the bare spaces following herbicide treatments to control broadleaf weeds. In conclusion, the UK Horse Pasture Evaluation Program has shown that educated horse farm managers better understand pasture dynamics and start to make management decisions based on knowledge rather than tradition.

[1]Krista.lea1@uky.edu, University of Kentucky.

INOCULANT EFFECTS ON FERMENTATION CHARACTERISTICS AND NUTRITIVE VALUE OF BERMUDAGRASS SILAGE

Joao Vendramini[1]

Our objective was to evaluate the effects of commercial silage inoculants on mycotoxins, fermentation characteristics, and nutritive value of bermudagrass silage. The experimental area was located at the Range Cattle Research and Education Center, Ona, FL. Plots (10 x 10 ft) of Jiggs bermudagrass were staged on September 15 2015 at 3 in. stubble height and fertilized with 80 lbs N/ac. Harvest occurred on October 15 2015 with target regrowth interval of 4 weeks. The inoculants tested were B500, Biotal Plus II, Early Sile Advance, Promote HQ, Promote VS-3, AS, and XC. Mini-silos (PVC pipes with capacity of 2 lb green forage) were filled after harvest with forage at 20-30% dry matter (DM). Inoculants were applied with a sprayer before ensiling. The silos were opened on February 19 2016. There were no effects of inoculant on silage nutritive value; mean values were: DM=25%, DM recovery=91%, CP=12.1%, NDF=68.1%, ADF=39.7%, IVTD=49%, and NDFD=25%. There were no effects of inoculant on fermentation; mean values were: pH=5.0, Lactic acid=2.0%, Acetic acid=1.4%, Propionic acid=0.5%, Butyric acid=0.3%, and Ammonia=19.3% CP. Aerobic stability (120 h) and mold and yeast counts were similar among treatments. There was no presence of aflatoxin (< 5 ppb), zearalenone (< 500 ppb) and T2 (< 500 ppb) on the silage in any treatment. Rresults differed from the first year of the research trial (2014) despite of similar forage characteristics in 2014 and 2015. There is a need to develop management practices to make the use of inoculants in warm-season grass silage more consistent and predictable.

[1]University of Florida, jv@ufl.edu.

AN EVALUATION INTO THE ROLE OF ENDOPHYTES IN SUMMER DORMANT TALL FESCUE

William Krogman, Erin Hubbard, Carolyn Young and Michael Trammell[1]

Summer active (SA) tall fescue is a successful forage grass that is widely used in the eastern United States because of its survivability, but much of it is toxic to grazing livestock because of an endophyte, called common toxic endophyte (CTE). Some SA tall fescue is now available with endophyte strains that are non-toxic but retain persistence. Summer dormant (SD) tall fescue could compliment traditional, annual small grains graze-out systems in this region by providing high quality forage from fall to spring. SD tall fescue differs from SA tall fescue by going dormant during summer, to avoid the heat and drought, offering multi-year persistence in the southern Great Plains. However, the role endophytes may play in SD tall fescue are not well understood. The objective of this study is to evaluate the role of endophytes in SD tall fescue to increase plant persistence. The study was conducted over two years by using clone pairs (endophyte-infected, E+ and endophyte-free, E-) that were generated from 10 tall fescue lines, two SA and eight SD. From 2015-2017, seed was harvested and agronomic data was collected on plant persistence. Bulked seed (lbs/line) showed no difference in the amount of seed produced between E+ and E-plants. Agronomic data from sward plots showed decreases in all tall fescue lines after two years of intensive grazing with SD tall fescue also displaying low cold tolerance. Future studies are looking at the east-west range in OK/TX and the north-south range in CA for SD tall fescue.

[1]Research Associate, Noble Research Institute, LLC., Ardmore, OK (wlkrogman@noble.org);
Research Associate, Noble Research Institute, LLC., Ardmore, OK (erhubbard@noble.org);
Associate Professor, Noble Research Institute, LLC., Ardmore, OK (cayoung@noble.org);
Senior Plant Breeder, Noble Research Institute, LLC., Ardmore, OK (matrammell@noble.org)

CURRENT BEST MANAGEMENT PRACTICES FOR HARVESTING AND STORING DRY HAY: A RESEARCH REVIEW

W.K. Coblentz[1]

The production of high-quality grass or legume hays in humid environments is complicated by slower drying rates, and increased probability of rainfall events compared to hay produced under arid climatic conditions. As a result, hay producers in humid environments often face the management dilemma of baling hay before it is adequately desiccated, or risking damage by rainfall events. Both options are undesirable, and result in reduced forage nutritive value. Forages damaged by rainfall events will lose sugars via leaching, or by prolonged or reactivated respiration. Damage is greater if rainfall events occur when forages are nearly dry enough to bale, and additional losses will occur during the additional field manipulation required to dry rain-damaged hay. As a result, fiber concentrations within the hay are increased, and the energy density of the forage is reduced. Some studies also suggest that a soaking rainfall event may reduce voluntary intake in cattle by about 10%, compared to undamaged hay. Forages baled before adequate desiccation has occurred are subject to spontaneous heating, which is regulated by bale moisture, bale size and/or density, climate, storage location/conditions, and the use of preservatives. Among these, the factor with the greatest impact is bale moisture. Bale size also can be particularly important, and becomes more problematic with large-round or large-square baling equipment, as well as with storage in large stacks. Consequences of spontaneous heating include respiratory losses of plant sugars, increased fiber components, formation of Maillard products (heat-damaged protein), reductions in energy density, and the potential for spontaneous combustion.

[1]Research Dairy Scientist/Agronomist, USDA-ARS, U.S. Dairy Forage Research Center, Marshfield, WI.

AUTHOR INDEX

AUTHOR INDEX

AUTHOR INDEX